D1228325

International Code
of
Nomenclature of Bacteria

BACTERIOLOGICAL CODE

International Code of Nomenclature of Bacteria

and

Statutes of the International Committee on Systematic Bacteriology

and

Statutes of the Bacteriology Section of The International Association of Microbiological Societies

BACTERIOLOGICAL CODE

1976 Revision

Approved by the Judicial Commission of the International Committee on Systematic Bacteriology, the International Committee on Systematic Bacteriology, the International Association of Microbiological Societies, and the Plenary Session of the 1st International Congress of Bacteriology, Jerusalem, Israel, September 1973

Edited by S. P. LAPAGE, P. H. A. SNEATH, E. F. LESSEL, V. B. D. SKERMAN, H. P. R. SEELIGER, and W. A. CLARK

Published for the International Association
of Microbiological Societies by the
American Society for Microbiology
Washington, D.C.
1975

Library of Congress Catalog Card Number 75-20730
ISBN 0-914826-04-2

American Society for Microbiology
Washington, D.C.
1913 I St., N.W.
Washington, D.C. 20006

CONTENTS

MEMORIAL TO PROFESSOR R. E. BUCHANAN

ROBERT E. BUCHANAN, 1883-1973

In his love for Latin and Greek, and of the etymology of names, Buchanan was a microbiologist extraordinary; that he had been an able administrator and an adviser to national and international bodies, a conservationist, and a benefactor of Iowa State University are aspects of his life that we, in this country, are apt to overlook.

Several members of the Buchanan family migrated from a village near Glasgow in the early 1800s; from New York they worked their way up the rivers to Chicago and were granted land in the state of Iowa. Buchanan's grandmother was a Chase whose family (best known for banking) went to America soon after the *Mayflower*. Robert Earle Buchanan was born in Cedar Rapids in 1883 and his interest in nature study was aroused at the age of nine, while attending a one-room country school near Rochester. Like most American boys, he worked during school holidays and saved to go to Iowa State College (ISC), which he entered in 1900. As a freshman, "Buchanan studied Latin under football coach Edgar Clinton" (Anon. 1965) and became a student laboratory assistant at 15 cents an hour to a botanist, L. H. Pammel, at ISC. He graduated in botany in 1904 and completed his master's degree in 1906. He spent some time in the medical school of the Northwestern University at Chicago and obtained his Ph.D. (majoring in bacteriology, with a minor in botany) in 1908.

At Iowa State College (ISC)

In 1910 Buchanan was appointed first head of bacteriology at ISC, and the same year married a botanist, Estelle Fogel, with whom he collaborated in writing the well-known Buchanan & Buchanan's *Bacteriology*. From 1914 to 1919 Buchanan was the first Dean of the Division of Industrial Sciences; from 1919 to 1948 he was the first Dean of the Graduate College, and from 1933 to 1948 Director of the Agricultural Experiment Station. When he retired officially in 1948 Buchanan was made 'Emeritus' and continued to have an office in the bacteriology department until his death; from this, and another office he had in Curtiss Hall, he kept a watchful eye on what went on in ISC, and he never hesitated to express his views forcibly when things displeased him. Throughout his life he took a great interest in Iowa State College (later University) and the Agricultural Experiment Station, and even after retirement his opinions were sought, respected, and sometimes feared.

In the summer vacations he would retire to the shores of Birch Lake in Minnesota, where he owned some land. There were two cabins (one belonged to his brother) built by their own hands, and over his boathouse Earle had a large office from which he sent a steady flow of dictaphone sleeves to his staff in Ames. His only relaxations were fishing and telling long tales of his travels, particularly of those in Arab countries.

In the cabin he was able to cook his fish by electricity (he was a good cook) for the cabin had all "mod. cons." except internal doors, for which curtains substituted.

Nearly twenty years after he retired, Iowa State University built and named after him a hall of residence for 400 graduate students.

The Scientific Side of the Dean

To his students Buchanan was always known as the Dean, and undoubtedly administration had been his forte in the prime of his life. It is hard to think of him working at a bench, but in 1918 he published a paper on the various phases of growing cultures. Most of his work was concerned with nomenclature and he was happiest delving into old books and holding forth about names. Between 1916 and 1918 he published a series of ten papers with the general title (subject to some variation) of "Studies in [on] the nomenclature and classification of [the] bacteria." In 1918 he was President of the SAB (Society of American Bacteriologists) and was a member of the Winslow Committee whose two reports (Winslow *et al.*, 1917, 1920) completely changed ideas on the classification and nomenclature of bacteria.

Of his other early publications, Buchanan's *General Systematic Bacteriology* (1925) is best known; it is a book of about 600 pages and gives a reasoned account of the names of bacterial genera and higher ranks. This book has become a classic and, because it is accurate and informative, it is still consulted.

International Committees and Congresses

In addition to being one of America's best-known bacteriologists at the age of 35, Buchanan became an international figure; he was sent by U.S. government departments and by FAO to several countries in the Middle East and to India to advise on agricultural matters. In a series of articles on past Presidents of the SAB, it was said of Buchanan that he was as well known a figure in Piccadilly as on the Ames campus.

In 1930 Buchanan presided over the bacteriology section of the Botanical Congress in Cambridge, and attended the first International Congress of Microbiology in Paris, where he became one of the founders of the

Nomenclature Committee. During the second Congress an American–Canadian Committee was set up to draft a code of bacteriological nomenclature and, of course, Buchanan became its chairman. He prepared a mimeographed document of 119 pages showing, in parallel columns, the International Rules of Botanical Nomenclature and the suggested wording for a bacteriological code based on the Botanical Code. A revised version was considered at the third Congress, when Buchanan was made the first chairman of the newly formed Judicial Commission. Further revision of the draft code followed and a Proposed Bacteriological Code (Buchanan and St. John-Brooks, 1947) was printed at Ames by ISC Press and circulated to members of the Nomenclature Committee for discussion at the fourth Congress. After amendment this Code was approved and published (Buchanan, St. John-Brooks, and Breed, 1948).

The object for which Buchanan had worked for so long had been achieved (or so it seemed) when an annotated version of the Code was published (Buchanan, Wikén, Cowan and Clark, 1958); the useful annotations were entirely Buchanan's work, though he insisted that the names of the other members of the Editorial Board should be included. Tinkering with the Code continued at each congress, for, like most editors, Buchanan could not forgo the pleasure of making alterations and amendments. At the end of the ninth Congress Buchanan resigned the chairmanship of the Judicial Commission, and he was made a Life Member of the Nomenclature Committee. The Society for General Microbiology made him an Honorary Member in 1957.

Buchanan established a unique position as the only person who attended all the International Congresses on Microbiology (of Microbiologists, of Microbiological Societies): 1930, Paris; 1936, London; 1939, New York; 1947, Copenhagen; 1950, Rio de Janeiro; 1953, Rome; 1958, Stockholm; 1962, Montreal; 1966, Moscow; 1970, Mexico City.

A minor but troublesome commitment Buchanan undertook was the setting up of an official publication for the Nomenclature Committee and the Judicial Commission. It had no financial backing but Buchanan secured help (a few hundred dollars) from UNESCO, encouragement from Iowa State College Press, and some printing from a small press about a hundred miles from Ames. But the world's most cumbersomely titled journal was born and later, with a glossy cover, achieved respectability as the *International Journal of Systematic Bacteriology*, which Buchanan edited until 1970.

Index Bergeyana and Bergey's Manual

After Bergey's death his *Manual* was carried on by R. S. Breed, E. G. D. Murray, and others, who made tentative plans for *Index Bergeyana*, an annotated list of names of bacterial taxa. Before this could be started Breed died, and Buchanan was invited to become Chairman of the Bergey's Manual Trust, an office he held until his death. All his life Buchanan had collected names of bacteria of all ranks and the record cards occupied a whole room in the office suite at Curtiss Hall; this collection became the major part of *Index Bergeyana* (Buchanan *et al.*, 1966) which could more appropriately have been named *Index Buchananensis.* There were many errors in the Index, some of fact, some of opinion on legitimacy, but it was a remarkable achievement. It was the work of a lifetime, but unfortunately it was published when the responsibilities of such a huge task pressed too heavily on an ageing man.

For the remaining years of his life preparations and plans for the eighth edition of *Bergey's Manual* occupied Buchanan's attention. He built up a team of strong-minded individualists who battled for several years with the problems leading to a new edition, and authors were chosen and invited to become contributors. Though he was interested primarily in the nomenclature, Buchanan never yielded a point and sometimes had authors and trustees tearing their hair at his insistence on a strict adherence to his beloved Code. With his attention focused on the names to be used in the *Manual,* his energies were dissipated on trivia; priority was always paramount, he was not concerned with usage or with the confusion that could arise when names were changed to conform with a strict application of the rules of nomenclature. As he aged and his judgments became less reliable, he became inconsistent and dogmatic; he found it difficult to understand numerical and computer approaches to bacterial classification, but this did not unduly concern him except when it might involve nomenclature, and then it might puzzle or even anger him.

Buchanan the Man

R. E. Buchanan was friendly, kind, and generous. As an American he was untypical, for even at his cabin in Minnesota he was formal and he never used Christian names when talking to or about colleagues. He was a man of strong character and liked to dominate a situation—and generally succeeded. His views were rigid and he was inflexible. In this he resembled Robert Breed, and when these two tough characters clashed the sparks would fly, often to the delight of the onlookers who took a less serious view of nomenclatural irregularities.

Buchanan could never understand why anyone should make light of his work, or be flippant about bacteriology, and worse, about its nomenclature. On one occasion he complained bitterly about the jocular attitude of Fred Bawden to virus names; he found Christopher Andrewes incomprehensible, for he, too, treated virus nomenclature in a cavalier manner. And, of course, he never saw the reasoning behind heretical taxonomy, which made its debut at a seminar in Ames.

In his intense interest in names and the meanings of words, and, during the later years of his life, an almost complete indifference to the biological aspects of bacteria, Buchanan was an unusual scientist. But without his uninhibited support for the importance of names, bacterial nomenclature will never be quite the same again.

S. T. COWAN
Queen Camel
Yeovil, England

REFERENCES

ANONYMOUS (1965). He looks to the future [a biographical note about Robert Earle Buchanan]. *News of Iowa State* **17**, no. 3 (pages not numbered).

BUCHANAN, R. E. (1918). Life phases in a bacterial culture. *Journal of Infectious Diseases* **23**, 109–125.

BUCHANAN, R. E. (1925). *General Systematic Bacteriology*. Baltimore: Williams and Wilkins.

BUCHANAN, R. E., HOLT, J. G. & LESSEL, E. F., JUN. (1966). *Index Bergeyana*. Baltimore: Williams and Wilkins.

BUCHANAN, R. E. & ST JOHN-BROOKS, R. (1947). *Proposed Bacteriological Code of Nomenclature*. Ames: Iowa State College Press. (Privately printed and of limited circulation.)

BUCHANAN, R. E., ST JOHN-BROOKS, R. & BREED, R. S. (1948). International bacteriological code of nomenclature. *Journal of Bacteriology* **55**, 287–306. Reprinted 1949, *Journal of General Microbiology* **3**, 444–462.

BUCHANAN, R. E., WIKÉN, T., COWAN, S. T. & CLARK, W. A. (1958). *International Code of Nomenclature of Bacteria and Viruses: Bacteriological Code*. Ames: Iowa State College Press.

WINSLOW, C.-E. A., BROADHURST, J., BUCHANAN, R. E., KRUMWIEDE, C., JUN., ROGERS, L. A. & SMITH, G. H. (1917). The families and genera of the bacteria: preliminary report of the Committee of the Society of American Bacteriologists on characterization and classification of bacterial types. *Journal of Bacteriology* **2**, 505–566.

WINSLOW, C.-E. A., BROADHURST, R. J. BUCHANAN, E., KRUMWIEDE, C., JUN., ROGERS, L. A. & SMITH, G. H. (1920). The families and genera of bacteria. Final report of the Committee of the Society of American Bacteriologists on characterization and classification of bacterial types. *Journal of Bacteriology* **5**, 191–229.

FOREWORD TO THE FIRST EDITION

Microbiologists who have occasion to use the scientific names of the microorganisms with which they deal generally prefer to use *correct* names and to use them *correctly*. Relatively few authors have special or direct interest in the problems of nomenclature as such, but there is general recognition that acceptance of the same names by various authors is essential in a field such as microbiology which has probably more economic implications than any other subdivision of biology. One is confronted with the fact that the names given to microorganisms have been proposed by individuals whose major interest has been the organisms themselves, not their names. Their economic significance has commonly been stressed. These minute organisms were found in some cases to produce disease in man, animals or plants; their study became basic to the professions of medicine and veterinary medicine; other microorganisms produced fermentation, decay and spoilage; it was found that fundamental studies of cellular physiology and metabolism, cell structure, inheritance, enzymology, photosynthesis, production of antibiotics, preservation of foods and feeds, public health, sanitation, soil fertility, plant pathology, and other fields required some basic knowledge of bacteriology. Those who discovered and worked with these organisms recognized the need of giving names to them, but frequently had little or no experience in scientific nomenclature. What rules should be followed in the coining of these names? Precedents to be followed were not clearly formulated in the early days of bacteriology.

Carl von Linné (Linnaeus) in the latter part of the eighteenth century proposed certain nomenclatural principles which were adopted with surprising unanimity by biologists of his day. Later the botanists and zoologists in separate international meetings and congresses developed two codes of nomenclature, which agreed in most points but differed in some. Many bacteriologists followed the Botanical Code, some the Zoological Code, and others named the organisms which they discovered with scant attention to established rules. It became evident that rules in Botany formulated primarily by those interested in the taxonomy of flowering plants, ferns and mosses did not fit too well the needs of the bacteriologist.

THE FIRST INTERNATIONAL MICROBIOLOGICAL CONGRESS (1930)

The desire that special attention should be paid to the peculiar needs of bacteriology was voiced at the First International Congress of Microbiology convened in Paris in 1930 by the International Society for Microbiology under the auspices of the Pasteur Institute. As the result of recommendations made by several of the delegates to the Congress, a Commission on Nomenclature and Taxonomy was constituted to prepare and report recommendations to the Plenary Session of the Congress.

The members of this commission were E. Pribram, Chicago, U.S.A., *Chairman*; A. R. Prévot, Paris, France, *Secretary*; R. E. Buchanan, Ames, Iowa, U.S.A.; K. Kisskalt, Germany; J. C. G. Ledingham, London, England; Reiner Müller, Köln, Germany; R. St. John-Brooks, London, England, and I. Yamasaki, Fukuoka, Kyushu, Japan.

Several resolutions prepared by the Commission were approved unanimously by the Plenary Session. These resolutions (in their English text) were as follows:

I. The founding of the International Society for Microbiology and the establishment of Congresses of Microbiology make possible for the first time adequate international cooperation relative to certain problems of microbial nomenclature. It is clearly recognized that the living forms with which the microbiologists concern themselves are in part plants, in part animals, and in part primitive. It is further recognized that *insofar as they may be applicable and appropriate* the nomenclatural codes agreed upon by international Congresses of Botany and Zoology should be followed in the naming of microorganisms. Bearing in mind however the peculiarly independent course of development that Bacteriology has taken in the past fifty years and elaboration of special descriptive criteria which bacteriologists have of necessity developed, it is the opinion of the International Society for Microbiology that the bacteria constitute a group for which special arrangements are necessary. Therefore, the International Society for Microbiology has decided to consider the subject of Bacterial Nomenclature as part of its permanent programme.

II. The International Society for Microbiology is of the opinion that the interests of bacterial nomenclature will best be served by placing the subject in the hands of a single International Committee, under the aegis of the International Society for Microbiology, adequately representative of all departments of Bacteriology, on which experts from all spheres of bacteriological research may work together. It is recognized that the subject of bacterial nomenclature is of so wide a nature that unless the personnel of an International Committee formed to deal with it is representative of all aspects of bacteriology, it is not likely to carry weight. Such a representative committee, to be called the Nomenclature Committee for the International Society for Microbiology, is hereby authorized and constituted.

III. The Nomenclature Committee for the International Society for Microbiology shall be constituted as follows:

a. Two permanent secretaries shall be elected: one primarily to represent medical and veterinary bacteriology, the other primarily to represent the other phases of bacteriology. The following individuals are hereby appointed secretaries.

(1) To represent primarily medical and veterinary bacteriology Dr. Ralph St. John-Brooks, Lister Institute, London, England.

(2) To represent primarily the other phases of bacteriology Dr. R. S. Breed, Geneva, New York, U.S.A.

Should a secretaryship become vacant, the position may be filled *pro tempore* by choice of the Committee. A permanent secretary should be chosen by action of the next succeeding International Congress for Microbiology.

b. The remaining members of the Committee shall be appointed by such National Committees of the International Society and by such of the various National Societies affiliated with the International Society as may desire representation thereon. Not more than three members may be thus chosen to represent a single nation. In addition, in order that the Committee shall be truly representative of all interests, the Committee is authorized to add such members as may be deemed desirable.

IV. The duties of the Nomenclature Committee shall include the following:

a. Through the secretaries the members of the Committee shall be circularized with reference to such problems of bacterial nomenclature as may arise, and shall endeavour to reach an agreement. No action relating to nomenclature shall be considered complete and operative until it has been considered by all members of the Committee, until adequate publicity has been given with respect to actions proposed, until approval has been given by a majority of two thirds of the members of the Committee, and until a report has been made to the next succeeding International Congress for Microbiology and opportunity thereby given for objection, modification or rejection by action of the Congress.

b. The Committee shall consider, among others, problems such as criteria to be employed in classification, adoption of names for *species* and *genera conservanda,* type species (including their identification and preservation), the encouragement of monographing of special groups or genera of bacteria by those best qualified to do the work, the enlargement of the scope and usefulness of the various type culture collections by more adequate support, and the preparation and publication of such Committee and Subcommittee reports as may be advisable.

V. Copies of these resolutions shall be submitted to the appropriate sections of the International Botanical Congress, Cambridge, 1930. It is the hope of the International Congress for Microbiology that the members of the International Botanical Congress who are interested in bacterial nomenclature will see the advisability of the special questions of nomenclature of bacteria being considered by a single international authority and that they will suggest names of members of the Botanical Congress willing to serve on the committee who, in their opinion, would add to its strength and authority.

VI. In view of the adequate provision made for special regulations relating to the bacteria, and the feasibility of designating *genera conservanda* among the bacteria by international agreement, it is believed that the greatest stability will be conferred by the adoption of the publication of *Species Plantarum* by Linnaeus in 1753 as the point of departure for bacterial nomenclature. The adoption of this date is recommended. It is further suggested that no present action be taken with reference to a list of *genera conservanda* for the bacteria.

VII. Among the most important agencies working toward satisfactory nomenclature and classification of bacteria are the several type culture collections. These constitute invaluable repositories and much of the future development of bacteriology will depend upon their adequate growth, support and utilization; in some cases at least they should develop into research institutes of high grade. It is urged that the coordination and cooperation existing among these institutions be extended the better to serve the interests of bacteriology in its theoretical, medical and other economic aspects. It is further urged that all bacteriologists publishing descriptions of new species or important strains of bacteria deposit pure cultures of such with a culture collection that they may be made available to others interested. Particularly is it urged that the adequate financial support of these culture collections by official agencies, by educational and research institutions and by the research foundations constitutes an important and immediate need.

It will be noted that in the action of the Congress the development of an adequate Bacteriological Code was linked with the Botanical Code. The specific suggestion was made that members of the International Botanical Congress, 1930, be apprised of the resolutions passed by the First Microbiological Congress and that the Botanical Congress be asked to cooperate. This was done, and the two secretaries of the International Nomenclature Committee for Bacteriology (Dr. R. St. John-Brooks and Dr. R. S. Breed) were designated by the Botanical Congress as a special committee on the nomenclature of bacteria.

THE SECOND INTERNATIONAL CONGRESS FOR MICROBIOLOGY (London, 1936)

The International Committee met during the sessions of the second International Congress for Microbiology in London in 1936. Proposals by R. E. Buchanan and H. J. Conn to conserve the generic name *Bacillus* Cohn 1872, to designate as the type species *Bacillus subtilis* Cohn 1872, and to fix the type or standard culture as the "Marburg strain" were approved by the Committee and by the Plenary Session of the Congress.

A further specific action of the Nomenclature Committee and of the London Congress had to do with the duplication of generic names in the *Protista,* the group ordinarily defined to include the protozoa, algae, fungi and bacteria. Inasmuch as bacteria are usually included among the plants, and subsequent plant homonyms are regarded as illegitimate, the principal interest is the suppression as illegitimate later homonyms in the protozoa and the bacteria. Prof. F. Mesnil proposed and the Nomenclature Committee and the Congress agreed that generic homonyms are not permitted in the group *Protista;* further that it is advisable to avoid homonymy amongst *Protista* on the one hand, plants or animals (*Metazoa*) on the other.

The Committee and Congress also acted favorably on a proposal by Prof. R. S. Breed relative to non-capitalization of specific epithets in names of species of bacteria.

> "Bacteriologists should accept Article 13 of the International Rules of Zoological Nomenclature, as follows:
>
> > 'While specific substantive names derived from names of persons may be written with a capital initial letter, all other specific names are to be written with a small initial letter.' "

At this 1936 (London) meeting of the International Committee it was agreed that, before the convening of the third International Congress of Microbiology to be held three years later in New York, a tentative Code of Bacteriological Nomenclature should be drafted and presented for the consideration of the Committee. To facilitate easy conference an American (Canadian and U.S.A.) Subcommittee was constituted to prepare such a tentative code. The members of this Subcommittee were R. E. Buchanan, Chairman; Robert S. Breed; J. Howard Brown; I. C. Hall; W. L. Holman; E. G. D. Murray; and Otto Rahn.

The chairman was asked to assemble material for consideration by the members. A mimeographed brochure of 119 pages was prepared under the title "Rules of Nomenclature, Annotated." It consisted of two parallel columns. In the first column the International Rules of Botanical Nomenclature, including *Principles, Rules, Recommendations, Notes* and *Examples* were printed. In the second column were listed suggestions for a code of Bacteriological Nomenclature formulated by making such minor modification of the Botanical Code as seemed desirable, as by dropping of inapplicable sections. In numerous footnotes were given the pertinent sections of the International Rules of Zoological Nomenclature and the American Code of Entomological Nomenclature. This material was sent to all members of the Subcommittee and to a large number of other bacteriologists, including members of the International Committee insofar as they could be reached. Criticisms and suggestions were invited. More than 30 sets of comments and suggestions were received. These comments were broken up into sections corresponding to those of the suggested code, and the proposed code and comments again submitted to the members of the Subcommittee in the form of a mimeographed booklet under the title "Suggestions and Comments on 'Rules of Nomenclature, Annotated'. " A new series of comments and suggestions was secured from the numerous collaborators, tabulated and submitted once more to the Subcommittee. A final revision was prepared to present to the International Committee at its New York meeting in 1939. The text of this tentative code differed from the basic Botanical Code principally in the following.

a. A reorganization of the text of the code under the following headings.
    1. General Considerations; 2. General Principles; 3. Rules of Bacteriological Nomenclature with Recommendations; 4. Provisions for interpretation and modification of rules.
b. Elimination of items and sections of the Botanical Code which seemed inapplicable to bacteriology.
c. Simplification where possible through rephrasing.
d. Selection of examples where possible from bacteriology.

THE THIRD INTERNATIONAL MICROBIOLOGICAL CONGRESS (New York, 1939)

The proposed tentative code was considered at some length by the International Committee for Bacteriological Nomenclature at its New York meeting; many suggestions developed. The report was also presented to one of the sections of the Congress, and about one hundred copies of the "Annotated" and "Tentative" codes distributed.

Upon recommendation of the International Committee on Bacteriological Nomenclature the Plenary Session of the Third International Congress for Microbiology on Sept. 9, 1939 approved the following resolution:

1. That a recognized Bacteriological Code be developed.
2. That publication of such a proposed Code when developed be authorized with the proviso that it shall be regarded as wholly tentative, but in the hope that it shall be widely tested so that it may be brought up for further consideration and final disposition at the next Microbiological Congress which should normally take place in 1942.
3. That the Nomenclature Committee, as at present constituted, shall continue to function under the auspices of the International Association of Microbiologists* as it did under the International Society for Microbiology.
4. That the International Committee shall select from its membership a Judicial Commission consisting of twelve members, exclusive of members *ex officio,* and shall designate a Chairman from the membership of the Commission. The two Permanent Secretaries of the International Committee on Bacteriological Nomenclature shall be members *ex officio* of the Judicial Commission. The commissioners shall serve in three classes of four commissioners each for nine years, so that one class of four commissioners shall retire at every International Congress. In case of resignation or death of any commissioner, his place shall be filled for the unexpired term by the International Committee at its next meeting.

The functions of the International Committee on Bacteriological Nomenclature were more accurately defined as follows:

a. To consider and pass upon all recommendations relating to the formation or modification of Rules of Nomenclature. The Com-

* The new name approved for the international organization sponsoring microbiological congresses.

mittee will recommend such action as may be appropriate to the next Plenary Session of an International Congress for Microbiology.

b. To consider all Opinions rendered by the Judicial Commission. Such Opinions become final if not rejected at the meeting of the International Committee next following the date on which the Opinion was issued.
c. To designate official Type Culture Collections.
d. To receive and act upon all reports and recommendations received from the Judicial Commission or other committees relating to problems of nomenclature or taxonomy.
e. To hold at least one meeting triennially in connection with the meeting of the International Congress for Microbiology.
f. To report to the final Plenary Session of each Congress a record of its actions, and to recommend for approval such actions as require the approval of the Congress.
g. To cooperate with other Committees, particularly those of the International Botanical and Zoological Congresses, to consider common problems of nomenclature.

The functions of the Judicial Commission of the International Committee on Bacteriological Nomenclature were also defined as follows:

a. To issue formal *Opinions* when asked to interpret rules of nomenclature in cases in which the application of a rule is doubtful.
b. To prepare formal *Opinions* relative to the status of names which have been proposed, placing such names when deemed necessary in special lists, such as lists of *Nomina Conservanda, Nomina Rejicienda,* etc.
c. To develop recommendations for emendations of the International Rules for Bacteriological Nomenclature for presentation to the International Committee.
d. To prepare formal *Opinions* relative to types, particularly types of species and genera, and to develop a list of bacterial genera which have been proposed with the type species of each.
e. To prepare and publish lists of names of genera which have been proposed for bacteria, for protozoa, or for other groups in which microbiologists are interested in order to assist authors of new names in avoiding illegitimate homonyms.
f. To develop a list of publications in microbiology whose names of organisms shall have no standing in bacteriology in determination of priority.
g. To edit and publish the International Rules of Bacteriological

Nomenclature, Opinions, Lists of *Nomina Conservanda, Nomina Rejicienda,* Type Species, etc.

h. To report to the International Committee at its triennial meetings all Recommendations, Transactions and Opinions.

i. To report to the International Committee at its triennial meetings the names of all Commissioners whose terms of service expire, likewise a list of all vacancies caused by resignation or death.

Recommendation. Whenever, in the opinion of any microbiologist an interpretation of any rule or recommendation is desirable because the correct application of such a rule or recommendation is doubtful, or the stability of nomenclature could be increased by the conservation or by the rejection of some name which is a source of confusion or error, it is recommended that he prepare a brief outlining the problem, citing pertinent references and indicating reasons for and against specific interpretations. This brief should be submitted to the Chairman of the Judicial Commission; if desired, through one of the Permanent Secretaries. An Opinion will be formulated, which may not be issued until it has been approved by at least eight members of the Commission.

It was further voted:

That the Proposed International Rules of Bacteriological Nomenclature, in so far as they have been developed by the American-Canadian Committee on Compilation of Proposals on Bacteriological Nomenclature for the International Committee and modified by action of that Committee, shall be referred for final emendation and publication to the Judicial Commission in accordance with Provision (c) above as recorded.

The minutes of the International Committee contain the following statements relative to the Judicial Commission:

With regard to the constitution of the Judicial Commission, members of the Committee present were requested to give its Secretaries lists of persons that they wished to nominate as members of the Judicial Commission, and the Secretaries were requested to transmit such nominations to the entire Committee for ballot, giving members the option of substituting other names if they so desired. It was agreed that after the final ballot the four persons receiving the greatest number of votes should be elected for the nine-year period and that the four persons receiving the smallest number of votes should be elected for the three-year period. The remaining four are to serve for a six-year period.

Nominations to membership on the Judicial Commission were made by the membership of the International Committee in attendance at the

New York meeting. The Permanent Secretaries then conducted a mail ballot resulting in the election of twelve members (Commissioners) and designation of R. E. Buchanan as Chairman. R. S. Breed and R. St. John-Brooks as Permanent Secretaries of the International Committee also became *ex officio* members and Permanent Secretaries of the Commission.

The records of the Congress showed a membership of 62 on the International Committee on Bacteriological Nomenclature as of August 1939. There were representatives of Microbiological Societies of 24 nations as follows: Argentina, Australia, Belgium, Brazil, Bulgaria, Canada, Denmark, Deutsches Reich, Eire, France, Great Britain, Holland, Hungary, Italy, Norway, Palestine, Poland, Roumania, Spain, Sweden, Switzerland, United States of America, Union of Soviet Socialist Republics, and Uruguay.

It was expected that the mandate of the Congress to the Judicial Commission to develop and publish a tentative proposal for a Code of Bacteriological Nomenclature would be followed promptly. The final determination of the constitution of the Judicial Commission itself was long delayed because of the outbreak of World War II while the New York Congress was in session. It soon proved impracticable to circulate copies of the nomenclatural proposals and to secure comments from all members of the Commission.

Dr. Ralph St. John-Brooks of the Lister Institute, London, one of the Permanent Secretaries of the International Committee in March 1942, spent some days with the Chairman of the Commission in conference and in editing the manuscript which had been reviewed by the Committee at the New York City meeting.

THE FOURTH INTERNATIONAL MICROBIOLOGICAL CONGRESS (Copenhagen, 1947)

The Proposed Bacteriological Code of Nomenclature as authorized by the Third International Congress for Microbiology was printed in June 1947 in a limited edition for distribution and for use by the Judicial Commission and the International Committee at the Fourth International Congress in September 1947.

At the Copenhagen Meeting the proposed Code was considered, revised, and approved for publication by the Judicial Commission, the International Committee and the Plenary Session of the Congress. The English text was published in March 1948 in the Journal of Bacteriology, later (Sept. 1949) in the Journal of General Microbiology. A Spanish translation (1949) by Prof. Verna was published in Argentina in De Archivos de Farmacia y Bioquimica del Tucumán and a German translation by

Dr. med. Hubert Bloch (1950) in the Schweizerische Zeitschrift für allgemeine Pathologie und Bakteriologie. A French translation by Dr. Prévot and a Japanese translation were also issued.

THE FIFTH INTERNATIONAL MICROBIOLOGICAL CONGRESS (Rio de Janeiro, 1950)

Meetings of the Judicial Commission and of the International Committee were held at Rio de Janeiro and Petropolis (Brazil) in August 1950. Among the important actions of these bodies, confirmed by the Plenary Session of the Congress, were the following:

1. An Editorial Board was established consisting of the Chairman of the Judicial Commission and the two Permanent Secretaries.
2. Publication of a quarterly "International Bulletin of Bacteriological Nomenclature and Taxonomy" was authorized; to be edited by the Editorial Board.
3. Agreement was reached that some revision of the International Bacteriological Code of Nomenclature was desirable and the Judicial Commission instructed to incorporate certain amendments approved, and to prepare recommendations for the 1953 International Microbiological Congress to be held in Rome.

The actions taken by the Commission, the Committee and the Fifth Congress are reported in Volume One of "The International Bulletin of Bacteriological Nomenclature and Taxonomy."

In preparation for the Rome Congress (September 1953), the provisional agenda for the meetings of the International Committee and of the Judicial Commission were prepared and published in the June (1953) issue of the International Bulletin.

THE SIXTH INTERNATIONAL MICROBIOLOGICAL CONGRESS (Rome, 1953)

The Editorial Board prepared a series of "Proposals Relative to Emendation and Publication" of a revised International Bacteriological Code of Nomenclature (International Bulletin 1953, pp. 31–63) which recommended publication of the revised Code, suggested that the Rules and Recommendations be adequately annotated, and that there be noted significant resemblances to the Botanical and Zoological Codes of Nomenclature and likewise important differences between them. The hope was expressed that texts in other languages could be published simultaneously with the English text. In all, sixty draft proposals for amendment, deletions and modifications of the Code were submitted and acted upon.

The Judicial Commission, through the Editorial Board, was directed to edit, annotate, and publish the Code as finally approved by the International Committee and the Plenary Session.

The name of the Code was fixed as The International Code of Nomenclature of Bacteria and Viruses.

The manuscript for the Code in original draft form, including Annotations and Appendices, was submitted for editorial suggestions to all members of the Judicial Commission and to about twenty-five bacteriologists experienced in nomenclature and taxonomy. Unfortunately the preparation of the text and annotations has been so time-consuming that it has not been possible to include texts of the Code in the several important languages of science. It is to be hoped that this may be done in future printings.

REFERENCES

BUCHANAN, R. E. 1939 (February) (Editor). Rules of Nomenclature: Annotated; with Suggestions for Rules of Bacteriological Nomenclature. Prepared for the American-Canadian Committee on Compilation of Proposals for Consideration by the Third International Congress for Microbiology. Mimeographed pp. 118. Ames, Iowa. U.S.A.

——— 1939 (July) (Editor). Proposed International Rules of Bacteriological Nomenclature. Prepared for Consideration by The American-Canadian Committee on Compilation of Proposals on Bacteriological Nomenclature for the International Committee. Mimeographed. pp. 63, Ames, Iowa. U.S.A.

———, and Ralph St. John-Brooks. 1947 (June) (Editors). Proposed Bacteriological Code of Nomenclature. Developed from proposals approved by International Committee on Bacteriological Nomenclature at the Meeting of the Third International Congress for Microbiology. Publication authorized in Plenary Session. pp. 61. Iowa State College Press, Ames, Iowa. U.S.A.

———, ——— and Robert S. Breed 1948 (March). (Editors). International Bacteriological Code of Nomenclature. Journ. Bact. 55:287–306. Also reprinted in September 1949 Journ. General Microbiology 3:444–462.

VERNA, LUIS C. (Translator) 1949. Codigo International de Nomenclatura Bacteriologica. Archivos de Farmacia y Bioquimica del Tucumán. 4:283–316. Tucumán, Argentina.

BLOCH, HUBERT (Translator) 1950. Internationaler bakteriologischer Nomenklaturcodex. Schweiz. Zeitschr. allgem. Path u. Bakteriologie. 13:358–383. Basel, Schweiz.

Editorial Board. 1953. The International Bacteriological Code of Nomenclature: Proposals relative to emendation and publication. Internatl. Bull. Bact. Nomenclature and Taxonomy 3:31–62. Iowa State College Press, Ames, Iowa. U.S.A.

COWAN, S. T. and T. WIKÉN (Secretaries) 1953. Minutes of the Judicial Commission Meetings held at Rome in Connection with the VI International Congress for Microbiology. September, 1953. Internatl. Bull. Bact. Nomenclature and Taxonomy 3:141–154. Iowa State College Press, Ames, Iowa. U.S.A.

——— 1953. Minutes of Meetings of the International Committee on Bacteriological Nomenclature held at Rome in Connection with the VI International Congress for Microbiology September, 1953. *Ibid.* 3:155–161. Iowa State College Press, Ames, Iowa. U.S.A.

SUPPLEMENT TO THE FOREWORD OF THE FIRST EDITION

The last book edition of the Code was that approved by the Sixth Congress, Rome, 1953 (International Code of Nomenclature of Bacteria and Viruses, 1958).

THE SEVENTH INTERNATIONAL CONGRESS FOR MICROBIOLOGY (Stockholm, 1958)

No changes to the Code were made at the Congress in Stockholm, and it was decided that matters pending should be presented to the Congress of 1962 (Cowan and Clark, 1958).

THE EIGHTH INTERNATIONAL MICROBIOLOGICAL CONGRESS (Montreal, 1962)

A large number of changes were made to the Code, mostly amplifications to cover problems that were arising in bacteriological nomenclature (Clark and Seeliger, 1963a, b). They concerned in particular the nomenclature of categories between genus and subgenus (Section, Subsection, Series, Subseries), recommendations on infrasubspecific names, generic descriptions, and citation and orthography. Many were taken with some modification from the Botanical Code. The amendments were published (Clark and Seeliger, 1963a) but a complete amended version of the Code was not published. Many of these changes were clearly necessary, but their insertion into the existing Code made it difficult to maintain a clear and logical order to the various rules.

THE NINTH INTERNATIONAL CONGRESS FOR MICROBIOLOGY (Moscow, 1966)

The Moscow Congress marked a change of direction in the philosophy of bacterial nomenclature. Change was in the air, and this is illustrated by the decision of the virologists (represented by the Subcommittee on Viruses of the International Committee on Nomenclature of Bacteria) to prepare their own rules of nomenclature. This led to the establishment at the Moscow Congress of a separate International Committee on Nomenclature of Viruses. This move was largely due to the feeling that viruses were of such a special nature that a new and different system of nomenclature should be introduced, and partly because Linnaean binary names were considered to be inappropriate (Cowan, 1963; Clark and Seeliger, 1967a, b). The first report of the Virus Committee was published in 1971 (Wildy, 1971).

At the same time the Executive Board of the International Association of Microbiological Societies requested all subordinate bodies to prepare and submit Statutes. In the first edition the statements covering the structure and functions of the International Committee on Nomenclature of Bacteria (ICNB) were contained in Provisions 4 and 5 of the Code. The Judicial Commission authorized the removal of these Provisions and the Executive Board of the ICNB proceeded with the formulation of Statutes.

At the Moscow Congress, the Judicial Commission was presented with a considerable list of proposed changes to the Bacteriological Code (Editorial Board, 1966; Clark and Seeliger, 1967a, b), of which the most lengthy were proposals to regulate the nomenclature of infrasubspecific forms, forms that had previously been subject only to recommendations on good practice. These proposals had, at Montreal, been deferred for further study, and it now became evident that they posed many difficulties that could not be avoided without consultation with epidemiologists, geneticists, biochemists, and others. These proposals were again referred back for further study.

The Commission discussed again the need for the regulation of names of sections, subsections, series and subseries. It became clear that these categories were used almost only within one genus, *Streptomyces*, whose taxonomy and nomenclature were increasingly at odds with modern practice in the rest of bacteriology. A feeling grew that it was a retrograde step to recognize complex rules for such categories if their need was diminishing, as awareness grew that many forms recognized as separate species of *Streptomyces* were more likely to be infrasubspecific variants. At its next meeting, the Commission agreed to remove from the Code the provisions that controlled the names of these categories, and this has been done in the present Code.

The revisions made at Moscow made it necessary to publish a new edition of the Code (International Code of Nomenclature of Bacteria, 1966).

Leicester Meeting of the Judicial Commission

It was decided to hold a special meeting of the Judicial Commission to consider a complete revision of the Code and some way of eliminating the thousands of forgotten and useless names. This meeting was held in Leicester in 1968 (Lessel, 1970), and the Judicial Commission quickly agreed that the Code needed a complete new version. Dr. S. P. Lapage offered to undertake a complete revision, and a Drafting Committee was set up consisting of W. A. Clark, S. P. Lapage (Chairman), E. F. Lessel, H. P. R. Seeliger, and P. H. A. Sneath to prepare a Revised Code, to embody the following: publication of names in a limited range of publications; obligate designation of types; methods on designation and the preservation of type strains; minimal descriptions of taxa; and alteration to the provisions for amending the Code in view of impending changes in the organization of the International Committee on Nomenclature of Bacteria.

The question of old and useless names was considered at length. The device used by the Zoological Code—whereby names disused for 50 years could be considered to be forgotten names (*nomina oblita*) and thereafter ignored—was not thought useful. There was the risk of discovering later that such a name had been used in this period, thus necessitating reinstatement. Another suggestion was that there should be block conservation of well-established names in certain publications of international repute. This had the disadvantage that much detailed taxonomic work would be required before such names could be conserved, and that there would be numerous appeals where the publications were perpetuating obvious errors.

The idea of a new starting date was then discussed. Similar suggestions had been raised in the past, but the important innovation was the proposal that an Approved List be prepared containing all names of taxa with current usage, and that at some given date in the future all other names should lose their standing in nomenclature. The Approved List would then be the basis for the nomenclature of the future. It was realized that the object of the change would be defeated if the old names were not available for re-use, because search of literature would still have to be made to avoid earlier homonyms, but on closer examination it was felt that the re-use of old names should not lead to major confusion. In the event, this radical proposal was accepted and is thought to be workable (Clark and Seeliger, 1971).

THE TENTH INTERNATIONAL CONGRESS FOR MICROBIOLOGY (Mexico City, 1970)

Only minor emendations (Lessel, 1971; Clark and Seeliger, 1971), mostly of an editorial nature, were made at Mexico to the Code that was currently in force (the Code as approved at Moscow). The International Committee also approved the Statutes and changed its name to the International Committee on Systematic Bacteriology.

The first drafts of the Revised Code were prepared by the Drafting Committee between 1968 and 1970, when two separate drafts were sent to the Judicial Commission, the second of which was discussed by the Judicial Commission at the tenth Congress. The draft was favourably received, so a resumé of the main changes that were proposed was presented to the International Committee on Nomenclature of Bacteria (Clark and Seeliger, 1971). The International Committee approved the main outline of the proposed Revised Code and later received copies of the fourth draft for comment. These comments were incorporated, and the fifth draft was published for comment in time for the next Congress at Jerusalem in 1973 (Lapage *et al.,* 1973).

FIRST INTERNATIONAL CONGRESS OF BACTERIOLOGY (Jerusalem, 1973)

The Revised Code as proposed (Lapage *et al.,* 1973) was approved by the Judicial Commission of the International Committee on Systematic Bacteriology and the Plenary Session of the First International Congress of Bacteriology, with minor amendments mostly editorial in nature (Lessel, 1974; Clark and Schubert, 1974), and its publication was authorized in book form in the present volume.

P. H. A. SNEATH
Leicester, England

REFERENCES

CLARK, W. A., and R. H. W. SCHUBERT. 1974. International Committee on Systematic Bacteriology. 1st International Congress for Bacteriology. Minutes of the Meetings, 2 and 6 September 1973. Binyanei Ha'ooma, Jerusalem, Israel. Int. J. Syst. Bacteriol. **24**:375–379.

CLARK, W. A., and H. P. R. SEELIGER. 1963a. Detailed minutes concerning actions taken on the emendation of the International Code of Nomenclature and Viruses during the meetings of the Judicial Commission of the International Committee on Bacteriological Nomenclature at the VIII International Microbiological Congress in Montreal, August, 1962. Int. Bull. Bacteriol. Nomen. Taxon. **13**:1–22.

CLARK, W. A., and H. P. R. SEELIGER. 1963b. Minutes of the first meeting of the International Committee on Bacteriological Nomenclature, Pathology Building, McGill University, Montreal, August 18, 1962. Int. Bull. Bacteriol. Nomen. Taxon. **13**:39–46.

CLARK, W. A., and H. P. R. SEELIGER. 1967a. Minutes of the Judicial Commission of the International Committee on the Nomenclature of Bacteria. Int. J. Syst. Bacteriol. **17**:59–72.

CLARK, W. A., and H. P. R. SEELIGER. 1967b. Minutes of the International Committee on Nomenclature of Bacteria meetings at the IX International Congress on Microbiology, Moscow, 1966. Int. J. Syst. Bacteriol. **17**:73–78.

CLARK, W. A., and H. P. R. SEELIGER. 1971. International Committee on Nomenclature of Bacteria, Tenth International Congress for Microbiology. Minutes of the Meetings 8 and 13 August 1970, Hotel Maria Isabel, Mexico City, Mexico. Int. J. Syst. Bacteriol. **21**:111–118.

COWAN, S. T. 1963. Request of the Virus Subcommittee. Int. Bull. Bacteriol. Nomen. Taxon. **13**:171–173.

COWAN, S. T., and W. A. CLARK. 1958. Minutes of the Meetings of the International Committee on Bacteriological Nomenclature held at Stockholm in connection with the VII International Congress for Microbiology, July–August, 1958. Int. Bull. Bacteriol. Nomen. Taxon. **8**:145–149.

Editorial Board. 1966. Proposed emendation of the International Code of Nomenclature of Bacteria and Viruses—with comments. Int. J. Syst. Bacteriol. **16**:341–369.

International Code of Nomenclature of Bacteria. 1966. Int. J. Syst. Bacteriol. **16**:459–490.

International Code of Nomenclature of Bacteria and Viruses. 1958. Iowa State College Press, Ames, Iowa, 186 pp.

LAPAGE, S. P., W. A. CLARK, E. F. LESSEL, H. P. R. SEELIGER, and P. H. A. SNEATH. 1973. Proposed Revision of the International Code of Nomenclature of Bacteria. Int. J. Syst. Bacteriol. **23**:83–108.

LESSEL, E. F. 1970. Judicial Commission of the International Committee on Nomenclature of Bacteria. Minutes of Meeting, September 1968, Leicester, England. Int. J. Syst. Bacteriol. **20**:1–8.

LESSEL, E. F. 1971. Minutes of the Judicial Commission of the International Committee on Nomenclature of Bacteria. Int. J. Syst. Bacteriol. **21**:100–103.

LESSEL, E. F. 1974. Judicial Commission of the International Committee on Systematic Bacteriology. Minutes of the Meeting, 29 August 1973, Jerusalem, Israel. Int. J. Syst. Bacteriol. **24**:379–380.

WILDY, P. 1971. Classification and nomenclature of viruses. First Report of the International Committee on Nomenclature of Viruses. S. Karger, Basel. (Monographs in Virology, Vol. 5, 81 pp.)

PREFACE TO THE FIRST EDITION

The history of the development of the 1958 Revised Edition of the International Code of Nomenclature of Bacteria and Viruses has been given in the Foreword. Here it is fitting that there be acknowledgement of the generous assistance given by many individuals and organizations in the preparation and editing of this Code.

The task of developing a wholly satisfactory Bacteriological Code is not complete. New problems involving nomenclature of the bacteria will arise and will require solutions. There have as yet been no final recommendations and no conclusions as to what special Rules and Recommendations will be needed to make functional any proposals to be made by the International Subcommittee on Taxonomy of the Viruses relative to virus nomenclature. The increasing use of terminologies applicable to strains and groups of bacteria of infrasubspecific rank makes necessary careful study of the best methods for preventing confusion, even some degree of nomenclatural chaos, in the naming of taxa of lower rank than subspecies. The growing recognition of the value of the type concept in standardization of names may mean the incorporation into the Code of a definition of Type Culture Collections and their functions in stabilization of bacteriological nomenclature.

A reading of the Annotations of the several Rules and Recommendations of the Bacteriological Code reveals a variance in terminology (sometimes in basic concepts) in the three Biological Codes of Nomenclature (Botanical, Zoological and Bacteriological). These differences have come about through the peculiarly independent development and history of Botany and of Zoology. The organization which can facilitate any attempt to reconcile these interdisciplinary differences must represent biology as a whole and on an international basis. The International Union of Biological Sciences would seem to be the agency able in some effective manner to develop fruitful consultations among the nomenclatural commissions of the three disciplines.

The Editorial Board and the Judicial Commission are most grateful for the generous subventions that have made possible publication of this revised Bacteriological Code. Organizations particularly helpful have been the International Union of Biological Sciences, the Society of American Bacteriologists, and the Society for General Microbiology. The Iowa State College has likewise been most generous in its provision of office facilities.

The Editorial Board is grateful also for permission given by the Commissions concerned to quote from the International Code of Botanical Nomenclature and from the International Code of Zoological Nomenclature where it has been desirable to compare resemblances and differences between these Codes and the text of the revised International Code of Nomenclature of Bacteria and Viruses. However, the final text of the International Code of Zoological Nomenclature had not been adopted in final form at the time of publication of the International Code of Nomenclature of Bacteria and Viruses (June, 1958). In consequence some quotations may not represent final action by the 1958 Zoological Congress. If there are here included unintentional misinterpretations, they will be corrected in later editions of the Bacteriological Code.

The manuscript for the Code in original draft form, including annotations and appendices, was submitted for editorial suggestions to all members of the Judicial Commission and to about thirty other bacteriologists experienced in nomenclature and taxonomy. The suggestions received were reviewed by the Judicial Commission. The Code represents a high degree of international cooperation. The Editorial Board wishes to express its real appreciation for the helpful cooperation received.

The Editorial Board

R. E. Buchanan, *Chairman*
T. Wikén, *Secretary*
(resigned 1 April 1957)

S. T. Cowan, *Secretary*
W. A. Clark, *Secretary*
(appointed 8 October 1957)

PREFACE TO THE PRESENT EDITION

This volume contains the edition of the *International Code of Nomenclature of Bacteria* approved by the Plenary Session of the First Congress for Bacteriology, Jerusalem, 1973. The volume also contains the *Lists of Conserved and Rejected Names of Bacterial Taxa* together with the *Opinions issued by the Judicial Commission,* and the *Statutes of the International Committee on Systematic Bacteriology* (ICSB), formerly the International Committee on Nomenclature of Bacteria (ICNB). These Statutes, which deal with the administration of the ICSB, were developed from Provisions 4 and 5 of the earlier Codes. The *Statutes of the Bacteriology Section of the International Association of Microbiological Societies* (IAMS) are also included.

A revision of the International Code of Nomenclature of Bacteria (1966) has been undertaken in an attempt to simplify the rules of nomenclature, thus encouraging wider use of the Code, and to provide a sound basis for bacterial systematics. This edition supersedes all previous editions of the International Code of Nomenclature of Bacteria.

To achieve these aims, certain principles were recently approved by the ICSB (Lessel, 1971), and these have been incorporated into the present edition.

A new starting date (1 January 1980 rather than 1 May 1753) for the nomenclature of bacteria is proposed so as to put into practice more meaningful requirements for the valid publication of names. New names and combinations must be published in the *International Journal of Systematic Bacteriology* (IJSB) or, if published previously elsewhere, an announcement of such publication must be made in the IJSB; a description or a reference to a previously and effectively published description of the named taxon must also be given in the IJSB and the type of a named taxon must be designated.

The ICSB is requesting its taxonomic subcommittees and other experts to propose lists of characteristics which will constitute the minimal standards for the description of various taxa. When these have been approved by the ICSB, the Code recommends that the description of each named taxon contain at least those characteristics specified in the minimal standards. In addition the Code recommends that, in the case of cultivable organisms, cultures of the type strains of newly named species and subspecies be deposited in culture collections from which they would be available.

For names published prior to 1 January 1980, Approved Lists of Bacterial Names will be compiled by the members of the taxonomic subcommittees and by other experts for approval by the Judicial Commission and the ICSB. Only the names of bacteria which are adequately described and for which there is a type or neotype strain, if the organism is cultivable, will be placed on the approved lists. In determinations of priority after 1 January 1980, then, only those names which appear on the approved lists of names or which are validated by publication in the IJSB after 1 January 1980 need be taken into consideration. Thus it will no longer be necessary to conduct extensive, frequently difficult literature searches merely for the purpose of determining the earliest name that was used for a bacterial taxon. Most important, however, will be the fact that after 1 January 1980 all of the validly published names for the bacteria will have clear and precise applications because the names will be associated with adequate descriptions and with type or neotype strains.

For this edition of the Code, the Drafting Committee prepared several revisions which were circulated to members of the Judicial Commission and to the ICSB for their comments. The work was begun in 1968, approved in principle by the Judicial Commission in 1970 (at the Xth International Congress of Microbiology, Mexico City), and culminated in publication as a proposed Revision in 1973 (Lapage *et al.*, 1973) for comment by the scientific community prior to presentation to the Judicial Commission, the ICSB, and the Plenary Session of the Bacteriology Section of IAMS at its Congress in Jerusalem, 1973. There, the published text was approved (with minor changes) and approval was also given for publication in book form of the text contained in this volume. The date on which this edition of the Code becomes effective is the date of publication of this volume.

Examples have been included in the Code where they were thought helpful to illustrate clauses, but in a few instances examples from bacteriology have not so far been found. These cases have been indicated, as the use of hypothetical examples or those taken from botany would appear to be misleading. In a few cases, however, hypothetical examples have been used to illustrate orthography in Appendix 9. On the authority of the Judicial Commission and the ICSB, some of the earlier Opinions of the Judicial Commission have been edited to remove minor inconsistencies.

A memorial to Professor R. E. Buchanan is included in the volume as a tribute to the debt that all microbiologists owe to him for the earlier editions of the International Code of Nomenclature of Bacteria and Viruses. We thank the editors of the *Journal of General Microbiology* and the Cambridge University Press for permission to reproduce the photograph and obituary to Professor Buchanan, which originally appeared in the *Journal of General Microbiology* (Cowan, 1973) and which is also published in this volume by courtesy of Dr. S. T. Cowan, whose work on bacterial nomenclature is widely appreciated.

It would not be possible to list all the many individuals who helped with the revision of the Bacteriological Code. Apart from the members of the Judicial Commission and ICSB whose many comments are gratefully acknowledged, we would especially like to thank Dr. S. T. Cowan, Dr. N. E. Gibbons, Professor Helen Heise, Mr. L. R. Hill, and Sir Graham Wilson for their help and advice. In particular we must mention Professor V. B. D. Skerman whose alternative versions provided us with much valuable material for passages of the text and for his help and advice throughout and, as Chairman of the ICSB, for his assistance in circulating copies of drafts and guiding this Code through the many problems that arose.

The Drafting Committee

S. P. Lapage, *Chairman Drafting Committee, and Editor for the International Code of Nomenclature of Bacteria*

P. H. A. Sneath, *Chairman, Judicial Commission*

E. F. Lessel, *Editor, International Journal of Systematic Bacteriology*

H. P. R. Seeliger, *Secretary for Subcommittees, International Committee on Systematic Bacteriology* now *President-Elect of the International Association of Microbiological Societies*

W. A. Clark, then *Executive Secretary, International Committee on Systematic Bacteriology*

January 1975

REFERENCES

Cowan, S. T. 1973. Obituary: Robert E. Buchanan, 1883–1973. J. Gen. Microbiol. 77:1–4.

International Code of Nomenclature of Bacteria. 1966. Int. J. Syst. Bacteriol. 16:459–490.

Lapage, S. P., W. A. Clark, E. F. Lessel, H. P. R. Seeliger, and P. H. A. Sneath. 1973. Proposed revision of the International Code of Nomenclature of Bacteria. Int. J. Syst. Bacteriol. 23:83–108.

Lessel, E. F. 1971. Minutes of the Judicial Commission of the International Committee on Nomenclature of Bacteria. Int. J. Syst. Bacteriol. 21:100–103.

Chapter 1

General Considerations

Chapter 1. GENERAL CONSIDERATIONS

General Consideration 1

The progress of bacteriology can be furthered by a precise system of nomenclature accepted by the majority of bacteriologists of all nations.

General Consideration 2

To achieve order in nomenclature, it is essential that scientific names be regulated by internationally accepted Rules.

General Consideration 3

The Rules which govern the scientific nomenclature used in the biological sciences are embodied in International Codes of Nomenclature (see Appendix 1 for a list of these Codes).

General Consideration 4

Rules of nomenclature do not govern the delimitation of taxa nor determine their relations. The Rules are primarily for assessing the correctness of the names applied to defined taxa; they also prescribe the procedures for creating and proposing new names.

General Consideration 5

This *Code of Nomenclature of Bacteria* applies to all bacteria. The nomenclature of certain other microbial groups is provided for by other Codes: fungi and algae by the Botanical Code, protozoa by the Zoological Code, and viruses by the Virological Code when it is approved (see Appendix 1).

General Consideration 6

This Code is divided into Principles, Rules, and Recommendations.

(1) The *Principles* (Chapter 2) form the basis of the Code, and the Rules and Recommendations are derived from them.

(2) The *Rules* (Chapter 3) are designed to make effective the Principles, to put the nomenclature of the past in order, and to provide for the nomenclature of the future.

(3) The *Recommendations* (Chapter 3) deal with subsidiary points and are appended to the Rules which they supplement. Recommendations do not have the force of Rules; they are intended to be guides to desirable practice in the future. Names contrary to a Recommendation cannot be rejected for this reason.

(4) Provisions for emendations of Rules, for special exceptions to Rules, and for interpretation of the Rules in doubtful cases have been made by the establishment of the International Committee on Systematic Bacteriology (ICSB) and its Judicial Commission, which acts on behalf of the ICSB (see Rule 1b and Statutes of the International Committee on Systematic Bacteriology, pp. 131–150 of this volume). The decisions of the Judicial Commission are not final until ratified by the ICSB, which procedure is to be understood throughout this Code but which is not always repeated at each mention. The official journal of the ICSB is the *International Journal of Systematic Bacteriology* (IJSB) formerly the *International Bulletin of Bacteriological Nomenclature and Taxonomy* (IBBNT). (Some other journal could be specified by the ICSB if required. Such possible future specification is implicit in the use of *"International Journal of Systematic Bacteriology"* or "IJSB" throughout this Code, but is not always repeated at each mention.)

(5) *Appendices* are added to assist in the application of this Code (see Contents).

(6) Reference is given in the *Index* to clauses in which Definitions of certain words used in the Code are provided. Such words are indicated in boldface type in the clause concerned and in the index, and they may be printed in boldface type elsewhere in this Code.

General Consideration 7

Nomenclature deals with the following:

(1) Terms used to denote the **taxonomic categories,** e.g., "species," "genus," and "family."

(2) Relative rank of the categories (see Rule 5).

(3) Names applied to individual taxa. A taxonomic group is referred to throughout this Code as a **taxon,** plural **taxa. "Taxonomic group"** is used in this Code to refer to any group of organisms treated as a named group in a formal taxonomy; it may or may not correspond to a category.

Examples. Name of a species, *Pseudomonas* (generic name) *aeruginosa* (specific epithet) ; name of a genus, *Pseudomonas;* name of a family, *Pseudomonadaceae;* name of an order, *Pseudomonadales.*

Chapter 2

Principles

Chapter 2. PRINCIPLES

Principle 1

The essential points in nomenclature are as follows.

(1) Aim at stability of names.

(2) Avoid or reject the use of names which may cause error or confusion.

(3) Avoid the useless creation of names.

Note. "**Name**" in this Code is used to refer to scientific names applied to bacteria (see Chapter 3, Section 3).

Principle 2

The nomenclature of bacteria is **independent** of botanical nomenclature, except for algae and fungi, and of zoological nomenclature, except protozoa. For these exceptions and for relationships with virological nomenclature, see General Consideration 5 and Rule 51b(4).

"**Independent**" means that the same name may be validly used for a taxon of bacteria as well as a taxon of plants or animals with the exceptions noted above.

Principle 3

The scientific names of all taxa are Latin or latinized words treated as Latin regardless of their origin. They are usually taken from Latin or Greek (see Chapter 3, Section 9 and Appendix 9).

Principle 4

The primary purpose of giving a name to a taxon is to supply a means of referring to it rather than to indicate the characters or the history of the taxon.

Principle 5

The application of the names of taxa is determined by means of nomenclatural types, referred to in this Code as types (see Chapter 3, Section 4).

Principle 6

The correct name of a taxon is based upon **valid publication, legitimacy,** and **priority of publication** (see Chapter 3, Section 5).

Principle 7

A name of a taxon has no status under the rules and no claim to recognition unless it is validly published (see Chapter 3, Section 5B).

Principle 8

Each order or taxon of a lower rank with a given **circumscription, position,** and **rank** can bear only one correct name, i.e., the earliest that is in accordance with the Rules of this Code. Provision has been made for exceptions to this Principle (see Rules 23a and 23b and the Statutes of the ICSB).

Note 1. The name of a species is a binary combination of generic name and specific epithet.

Note 2. (i) by **circumscription** is meant an indication of the limits of a taxon, (ii) by **position** is meant the higher taxon in which a taxon is placed when there may be alternatives (see also Rule 23a), and (iii) by **rank** is meant level in the hierarchical sequence of taxonomic categories.

Principle 9

The name of a taxon should not be changed without sufficient reason based either on further taxonomic studies or on the necessity of giving up a nomenclature that is contrary to the Rules of this Code.

Chapter 3

Rules of Nomenclature with Recommendations

Chapter 3. RULES OF NOMENCLATURE WITH RECOMMENDATIONS

Section 1. General

Rule 1a

This revision supersedes all previous editions of the *International Code of Nomenclature of Bacteria* (see Appendix 1). It shall be cited as *Bacteriological Code* (1975 Revision) and will apply from the date of publication (1 January 1976).

Rule 1b

Alterations to this Code can only be made by the ICSB at one of its plenary sessions. Proposals for modifications should be made to the Editorial Secretary in sufficient time to allow publication in the IJSB before the next International Congress of Bacteriology. For this and other Provisions, see the Statutes of the ICSB, pp. 131–150.

Rule 2

The Rules of this Code are retroactive, except where exceptions are specified.

Rule 3

Names contrary to a Rule cannot be maintained, except that the International Committee on Systematic Bacteriology, on the recommendation of the Judicial Commission, may make exceptions to the Rules (see Rule 23a and the Statutes of the ICSB).

Rule 4

In the absence of a relevant Rule or where the consequences of a Rule are uncertain, a summary in which all pertinent facts are outlined should be submitted to the Judicial Commission for consideration (see Appendix 8 for preparation of a Request for an Opinion).

Section 2. Ranks of Taxa

Rule 5a

Definitions of the taxonomic categories will inevitably vary with individual opinion, but the relative order of these categories may not be altered in any classification.

Rule 5b

The taxonomic categories above and including species which are covered by these Rules are given below in ascending taxonomic rank. Those in the left-hand column should be recognized where pertinent; those in the right-hand column are optional. The Latin equivalents are given in parentheses.

Species (*Species*)	
	Subgenus (*Subgenus*)
Genus (*Genus*)	
	Subtribe (*Subtribus*)
	Tribe (*Tribus*)
	Subfamily (*Subfamilia*)
Family (*Familia*)	
	Suborder (*Subordo*)
Order (*Ordo*)	
	Subclass (*Subclassis*)
Class (*Classis*)	

Rule 5c

A species may be divided into subspecies, which are dealt with by the Rules of this Code. **Variety** is a synonym of subspecies; its use is not encouraged as it leads to confusion, and after publication of this Code the use of variety for new names will have no standing in nomenclature.

Rule 5d

Taxa below the rank of subspecies (**infrasubspecific subdivisions**) are not covered by the Rules of this Code, but see Appendix 10.

Section 3. Naming of Taxa

General

Rule 6

The scientific names of all taxa must be treated as Latin; names of taxa above the rank of species are single words.

Recommendation 6

To form new bacterial names and epithets, authors are advised as follows.

(1) Avoid names or epithets that are very long or difficult to pronounce.

(2) Make names or epithets that have an agreeable form that is easy to pronounce when latinized.

(3) Avoid combining words from different languages, **hybrid names** (*nomina hybrida*).

(4) Not to adopt unpublished names or epithets found in authors' notes, attributing them to the authors of such notes, unless these authors have approved publication.

(5) Give the etymology of new generic names and of new epithets.

(6) Determine that the name or epithet which they propose is in accordance with the Rules.

Names of Taxa above the Rank of Genus

Rule 7

The name of a taxon above the rank of genus is a substantive or an adjective used as a substantive of Latin or Greek origin, or a latinized word. It is in the feminine gender, the plural number, and written with an initial capital letter.

Example: Family *Pseudomonadaceae.*

Historically, all these names were feminine plural adjectives qualifying the word "*plantae,*" plants; in modern bacterial nomenclature they qualify the word "*procaryotae.*"

Example: *Plantae pseudomonadaceae; Procaryotae pseudomonadaceae.*

In practice, such names are used alone and as substantives.

Example: A member of the *Pseudomonadaceae.*

Names of Taxa above the Rank of Order

Rule 8

The name of each taxon above the rank of order is a Latin or latinized word preferably in conformity with Recommendation 6. It is based by choice on a combination of characters of the taxon or from a single character of outstanding importance.

Examples: Kingdom–*Procaryotae;* Class–*Schizomycetes.*

Names of Taxa between Subclass and Genus (Order, Suborder, Family, Subfamily, Tribe, Subtribe)

Rule 9

The name of a taxon between subclass and genus is formed by the addition of the appropriate suffix to the stem of the name of the type genus (see Rule 15). These suffixes are as follows:

TABLE 1. *Suffixes for Categories*

Rank	Suffix	Example
Order	*-ales*	*Pseudomonadales*
Suborder	*-ineae*	*Pseudomonadineae*
Family	*-aceae*	*Pseudomonadaceae*
Subfamily	*-oideae*	*Pseudomonadoideae*
Tribe	*-eae*	*Pseudomonadeae*
Subtribe	*-inae*	*Pseudomonadinae*

Names of Genera and Subgenera

Rule 10a

The name of a genus or subgenus is a substantive, or an adjective used as a substantive, in the singular number and written with an initial capital letter. The name may be taken from any source and may even be composed in an arbitrary manner. It is treated as a Latin substantive.

Examples: Single Greek stem, *Clostridium*; two Greek stems, *Haemophilus*; single Latin stem, *Spirillum*; two Latin stems, *Lactobacillus*; hybrid name, Latin-Greek stems, *Flavobacterium*; latinized personal name, *Shigella*; arbitrary name, *Ricolesia.*

Recommendation 10a

The following Recommendations apply when forming new generic or subgeneric names.

(1) Refrain from naming genera and subgenera after persons quite unconnected with bacteriology or at least with natural science.

(2) Give a feminine form to all personal generic and subgeneric names whether they commemorate a man or a woman.

(3) Avoid introducing into bacteriology as generic names such names as are in use in botany or zoology, in particular well known names.

Rule 10b

Generic and subgeneric names are subject to the same Rules and Recommendations, except that Rule 10c applies only to subgeneric names.

Rule 10c

The name of the subgenus, when included in the name of a species, is placed in parentheses between the generic name and specific epithet.

Example: *Bacillus (Bacillus) subtilis; Bacillus (Aerobacillus) polymyxa.*

Names of Taxa between Subgenus and Species

Rule 11

The taxonomic categories *section, subsection, series,* and *subseries* are informal categories not regulated by the Rules of this Code. Their designations do not compete with the names of genera and subgenera as to **priority** and **homonymy.**

Note. **Priority** (see Chapter 3, Section 5A) means that the name or epithet first published in accordance with the Rules is the correct name, or epithet, for a taxon (see Rule 23a). **Homonymy** is the term applied when the same name is given to two or more different taxa of the same rank based on different types. The first published name is known as the **senior homonym** and any later published name as a **junior homonym.**

Names of Species

Rule 12a

The name of a species is a **binary combination** consisting of the name of the genus followed by a single **specific epithet.**

If a specific epithet is formed from two or more words, then the words must refer to a single concept and are to be joined. If the words were not joined in the original publication, then the epithet is not to be rejected but the form is to be corrected by joining the words, which can be done by any author. If an epithet has been hyphenated, its parts should be joined. The name retains its validity and standing in nomenclature.

Example: *Salmonella typhi murium* should be corrected to *Salmonella typhimurium.*

Rule 12b

No specific or subspecific epithets within the same genus may be the same if based on different types (see Rule 13c and Chapter 3, Section 9).

Example: *Corynebacterium helvolum* (Zimmermann 1890) Kisskalt and Berend 1918 is based on the type of *Bacillus helvolus* Zimmermann 1890; the specific epithet *helvolum* cannot be used for *Corynebacterium helvolum* Jensen 1934, another bacterium whose name is based on a different type.

Rule 12c

A specific epithet may be taken from any source and may even be composed arbitrarily.

Example: *etousae* in *Shigella etousae* derived from European Theater of Operations of the U.S. Army.

A specific epithet must be treated in one of the three following ways.

(1) As an adjective that must agree in gender with the generic name.

Example: *aureus* in *Staphylococcus aureus*.

(2) As a substantive in apposition in the nominative case.

Example: *radicicola* in *Bacillus radicicola*.

(3) As a substantive in the genitive case.

Example: *lathyri* in *Erwinia lathyri*.

Recommendation 12c

Authors should attend to the following Recommendations, and those of Recommendation 6, when forming specific epithets.

(1) Choose a specific epithet that, in general, gives some indication of a property or of the source of the species.

(2) Avoid those that express a character common to all, or nearly all, the species of a genus.

(3) If taken from the name of a person, it should recall the name of one who discovered or described it, or was in some way connected with it, and possess the appropriate gender (see Appendix 9B).

(4) Avoid in the same genus epithets which are very much alike, especially those that differ only in their last letters.

(5) Avoid the use of the genitive and the adjectival forms of the same specific epithet to refer to two different species of the same genus (see Rule 63).

Names of Subspecies

Rule 13a

The name of a subspecies is a **ternary combination** consisting of the name of a genus followed by a specific epithet, the abbreviation "subsp." (*subspecies*), and finally the **subspecific epithet.**

Example: *Bacillus cereus* subsp. *mycoides* (Flügge 1886) Smith *et al.* 1946.

For **"variety"** see Rule 5c.

Rule 13b

A subspecific epithet is formed in the same way as a specific epithet. When adjectival in form, it agrees in gender with the generic name.

Rule 13c

No two subspecies within the same species or within the same genus may bear the same subspecific epithet (see also Rule 12b).

Names of Infrasubspecific Subdivisions

Rule 14a

The designations of the various taxa below the rank of subspecies are not subject to the Rules and Recommendations of this Code. (For advice on their nomenclature see Appendix 10.)

Rule 14b

A Latin or latinized infrasubspecific designation may be elevated by a subsequent author to the status of a subspecies or species name providing that the resulting name is in conformity with the Rules. If so elevated, it ranks for purposes of priority from its date of elevation and is attributed to the author by whom it was elevated, provided that the author who elevates it observes Rule 27.

Example: *Leptospira javanica* Collier 1948, elevation of *Leptospira* sp. serovar *javanica* of Essevald and Mochtar 1938 by Collier in 1948.

Section 4. Nomenclatural Types and Their Designation

General

Rule 15

A taxon consists of one or more elements. For each named taxon of the various taxonomic categories (listed below), there shall be designated a **nomenclatural type.** The nomenclatural type, referred to in this Code as **"type,"** is that element of the taxon with which the name is permanently associated. The nomenclatural type is not necessarily the most typical or representative element of the taxon. The types are dealt with in Rules 16-22.

Types of the various taxonomic categories can be summarized as follows:

TABLE 2. *Taxonomic Categories*

Taxonomic category	Type
Subspecies Species	Designated strain; in special cases the place of the type strain may be taken by a description, preserved specimen, or an illustration
Subgenus Genus	Designated species
Subtribe Tribe Subfamily Family Suborder Order	Genus on whose name the name of the higher taxon is based
Subclass Class	One of the contained orders

Rule 16

After the date of publication of this Code, the type of a taxon must be designated by the author at the time the name of the taxon is proposed.

Rule 17

The type determines the application of the name of a taxon if the taxon is subsequently divided or united with another taxon.

Example: If the genus *Bacillus* is divided into the genera *Bacillus* and *Aerobacillus,* the genus which contains the type species, *Bacillus subtilis,* must be named *Bacillus.*

Type of a Species or Subspecies

Rule 18a

Whenever possible the type of a species or subspecies is a designated strain.

A type strain is made up of living cultures of an organism which are descended from a strain designated (except as in Rule 18c) as the nomenclatural type. The strain should have been maintained in pure culture and should agree closely in its characters with those in the original description (see Chapter 4, C). The type strain may be designated in various ways (see Rules 18b, c, d, e, and f).

For a species which has not so far been maintained in laboratory culture or for which a type strain does not exist, a description, preserved specimen, or an illustration (see also Rule 18h) may serve as the type.

Example: Non-cultivated, *Oscillospira guilliermondi* Chatton and Perard 1913.

Rule 18b *Designation by original author*

If the author in the original publication of the name of a species or subspecies definitely designated a type strain, then this strain shall be accepted as the type strain and may be referred to as the **holotype.**

Rule 18c *Designation by monotypy*

If the original author described only a single strain and did not designate it as the holotype, then this strain shall be accepted as the type strain and may be referred to as the **monotype.**

Rule 18d *Designation as lectotype*

If no holotype exists, one of the strains on which the original author based his description may be later designated as the type strain by the original or a subsequent author, and this strain shall be accepted as the type strain and may be referred to as the **lectotype.**

Rule 18e *Designation as neotype*

If a strain on which the original description was based cannot be found, a **neotype** strain may be proposed.

A **neotype** strain must be proposed (**proposed neotype**) in the IJSB, together with citation of the author(s) of the name, a description or reference to an effectively published description, and a record of the permanently established culture collection(s) where the strain is deposited (see also Note 1 to Rule 24a).

The author should show that a careful search for the strains used in the original description has been made and that none of them can be found. The author should also demonstrate that the proposed neotype agrees closely with the description given by the original author.

The neotype becomes established (**established neotype**) two years after the date of its publication in the IJSB, provided that there are no objections, which must be referred within the first year of the publication of the neotype to the Judicial Commission for consideration.

Rule 18f

A strain suggested as a neotype but not formally proposed in accordance with the requirements of Rule 18e (suggested neotype) has no standing in nomenclature until formally proposed and established.

Rule 18g

If an original strain that should constitute the type of a species is discovered subsequent to the formal proposal or establishment of a neotype for that species, the matter shall be referred immediately to the Judicial Commission.

Rule 18h

If a description or illustration constitutes, or a dead preserved specimen has been designated as the type of a species (Rule 18a, para 3) and later a strain of this species is cultivated, then the type strain may be designated by the person who isolated the strain or by a subsequent author. This type strain shall then replace the description, illustration, or preserved specimen as the nomenclatural type.

Rule 18i *Change in characters of type and neotype strains*

If a type or neotype strain has become unsuitable due to changes in its characters or for other reasons, then the matter should be referred to the Judicial Commission, which may decide to take action leading to replacement of the strain.

Rule 19 *Reference strains*

A **reference strain** is a strain that is neither a type nor a neotype strain but a strain used in comparative studies, e.g., taxonomic or serological, or for chemical assay.

A **reference strain** has no standing in nomenclature, but it may, by subsequent action, be made a neotype.

Type of a Genus

Rule 20a

The nomenclatural type of a genus or subgenus is the type species, that is the single species or one of the species included when the name was originally validly published.

Rule 20b *Designation by original author*

If the author of the original publication of a generic or subgeneric name designated a type species, that species shall be accepted as the type species.

Rule 20c *Genus with only one species*

If the genus when originally published included only one species, then that species is the type species.

Rule 20d *Designation by a subsequent author*

The type species shall be selected from one of the species included when the genus was originally published.

Recommendation 20d

Authors are recommended to exclude the following species from consideration in selecting the type.

(1) Doubtfully identified or inadequately characterized species.

Example: *Lactobacillus caucasicus* Beijerinck 1901 (Opinion 38).

(2) Species doubtfully referred to the genus.

Example: No example yet found.

(3) Species which definitely disagree with the generic description.

Example: *Halococcus litoralis* (Poulsen) Schoop 1935.

(4) Species mentioned as in any way exceptional, including species which possess characters stated in the generic description as rare or unusual.

Example: *Pseudomonas mallei* (Zopf) Redfearn *et al.* 1966.

Rule 20e *Designation by international agreement*

(1) If none of the species named by an author in the original publication of a generic name can be recognized, i.e., if no identifiable type species can be selected in accordance with the Rules, the Judicial Commission may issue an Opinion declaring such generic name to be a **rejected name** (*nomen rejiciendum*) and without standing in nomenclature (see Rule 23a, Note 4).

Example: Rejection of the generic name *Gaffkya* Trevisan 1885 (Opinion 39).

(2) However, a generic name for which no identifiable type species can be selected in accordance with the Rules might have come into use for identifiable species which were subsequently named. In this case, one of these later species may be selected as the type species and established as such by an Opinion of the Judicial Commission. The generic name is then ascribed to the author of the name of the species selected as the type species.

Example: *Vibrio* Pacini 1854 and its type species *Vibrio cholerae* Pacini 1854 (Opinion 31).

Rule 20f *Retention of type species on publication of a new generic name*

The publication of a new generic name as a deliberate substitute for an earlier one does not change the type species of the genus.

Example: The deliberate creation of *Xanthomonas* as a substitute for the name *Phytomonas* (not available, as it was already in use as the name of a protozoan genus) does not change the type species, which was *Phytomonas campestris* and which became *Xanthomonas campestris.*

Type of a Subgenus

Rule 20g

A genus and its type subgenus share the same type species.

Example: *Bacillus subtilis* is the type species of the genus *Bacillus* and of its type subgenus, *Bacillus.*

Type of a Taxon from Genus to Order
(Subtribe, Tribe, Subfamily, Family, Suborder, and Order)

Rule 21a

The nomenclatural type of a taxon above genus, up to and including order, is the genus on whose name the name of the relevant taxon is based. One taxon of each category must include the type genus. The names of the taxa which include the type genus must be formed by the addition of the appropriate suffix to the stem of the name of the type genus (see Rule 9).

Example: Order, *Rhodospirillales*; suborder, *Rhodospirillineae*; family, *Rhodospirillaceae*; type genus, *Rhodospirillum.*

Rule 21b

If the name of a family was not made in conformity with Rule 21a but its name has been conserved, then the type genus may be fixed by an Opinion of the Judicial Commission.

Example: The genus *Escherichia* is the type genus of the family *Enterobacteriaceae* (Opinion 15).

Type of a Taxon Higher than Order

Rule 22

The type of a taxon higher than order is one of the contained orders, and if there is only one order this becomes the type. If there are two or more orders the type shall be designated by the author at the time of the proposal of the name.

Example: The order *Mycoplasmatales* of the class *Mollicutes.*

If not designated, the type of a taxon higher than order may be later designated by an Opinion of the Judicial Commission.

Example: None of the Opinions so far issued (1-51) has dealt with this subject.

Section 5. Priority and Publication of Names

A. Priority of Names

Rule 23a

Each taxon above species, up to and including order, with a given circumscription, position, and rank can bear only one correct name, that is, the earliest that is in accordance with the Rules of this Code.

The name of a species is a binary combination of generic name and specific epithet (see Rule 12a). In a given **position,** a species can bear only one correct epithet, that is, the earliest that is in accordance with the Rules of this Code.

Example: The species *Haemophilus pertussis* bears this name in the genus *Haemophilus*. If placed in the genus *Bordetella,* it bears the name *Bordetella pertussis.*

Note 1. In the case of a species, Rule 23a must be applied independently to the generic name and the specific epithet. The specific epithet remains the same on transfer of a species from one genus to another unless the specific epithet has been previously used in the name of another species or subspecies in the genus to which the species is to be transferred (see Rule 41a).

Note 2. The name of a subspecies is a ternary combination of a generic name, a specific epithet, and a subspecific epithet (see Rule 13c). In a given position a subspecies can bear only one correct subspecific epithet, that is, the earliest that is in accordance with the Rules of this Code. In the case of a subspecies, Rule 23a must be applied independently to the specific and subspecific epithets. The subspecific epithet remains the same on transfer of a subspecies from one species to another, unless the subspecific epithet has been previously used in the name of another species or subspecies in the genus to which the subspecies is to be transferred (see Rule 41a).

Note 3. The date from which all priorities were determined under the previous editions of the Code was 1 May 1753. After 1 January 1980, under Rule 24a all priorities will date from 1 January 1980 (see also Rule 24b).

Note 4. The Judicial Commission may make exceptions to Rule 23a by the addition of names to the list of **conserved names** (*nomina conservanda*) or to the list of **rejected names** (*nomina rejicienda*) (see Appendix 4). The Judicial Commission may correct the approved lists (see Rule 24a).

(i) By **conserved name** (*nomen conservandum*) is meant a name which must be used instead of all earlier **synonyms** and **homonyms.** By rejected name (*nomen rejiciendum*) is meant a name which must not be used to designate any taxon. Only the Judicial Commission can conserve or reject names (see also Rules 56a, b).

(ii) **Opinions** on the conservation or rejection of names, issued by the Judicial Commission, are published with other Opinions in the IJSB. Opinions are now numbered serially.

Note 5. Names and epithets may be:

legitimate—in accordance with the Rules;

illegitimate—contrary to the Rules;

effectively published—in printed matter made generally available to the scientific community (see Rule 25);

validly published—effectively published and accompanied by a description of the taxon or a reference to a description and certain other requirements (see Rules 27-32);

correct—the name which must be adopted for a taxon under the Rules.

Rule 23b

The date of a name or epithet is that of its valid publication. For purposes of priority, however, only legitimate names and epithets are taken into consideration (see Rules 32b and 54).

B. Publication of Names

Rule 24a

Valid publication of names (or epithets) which are in accordance with the Rules of this Code dates from the date of publication of the Code.

After 1 January 1980, priority of publication shall date from 1 January 1980. On that date all names published prior to 1 January 1980 and included in the Approved Lists of Bacterial Names of the ICSB shall be treated for all nomenclatural purposes as though they had been validly published for the first time on that date, the existing types being retained (but see Rule 24b).

Note 1. Names of bacteria in the various taxonomic categories validly published up to 31 December 1977 will be assessed by the Judicial Commission with the assistance of taxonomic experts. Lists of names will be prepared together with the names of the authors who originally proposed the names. When approved by the ICSB these Approved Lists of Bacterial Names will be published in the IJSB before 1 January 1980. Names validly published under this Code between 1 January 1978 and 1 January 1980 will be added to the Approved Lists of Bacterial Names.

After 1 January 1980, no further names will be added to the lists. Those names validly published prior to 1 January 1980 but not included in the approved lists will have no further standing in nomenclature. They will not be added to the lists of *nomina rejicienda* and will thus be available for reuse for the naming of new taxa.

The Approved Lists of Bacterial Names will contain for each name a reference to an effectively published description and the type whenever possible. In the case of species or subspecies, if a type strain is available, it will be listed by its designation, and the culture collection(s) from which it may be obtained will be indicated. If such a strain is not available, a reference strain or reference material will be listed if possible. Neotypes may be proposed in conformity with Rule 18e on such lists. (For citation of names on the approved lists, see Provisional Rules A1 and A2, pp. 33–34.)

Note 2. These approved lists may contain more than one name attached to the same type **(objective synonyms)** since the names on the lists will represent those names which are considered reasonable in the present state of bacteriological nomenclature and taxonomy and represent the views of many bacteriologists who may hold different taxonomic opinions.

Note 3. Synonyms may be **objective synonyms,** i.e., more than one name has been associated with the same type, or **subjective synonyms,** i.e., different names have been associated with different types that in the opinion of the bacteriologist concerned belong to the same taxon. The synonym first published is known as the **senior synonym,** and later synonyms are known as **junior synonyms.**

Publication of **objective synonyms** in the lists does not affect bacterial nomenclature any more than does the valid publication of objective synonyms in different works in the bacteriological literature at present.

Examples: **Objective synonyms**—*Nocardia rhodochrous* and *Mycobacterium rhodochrous.* **Subjective synonyms**—Luedemann (IJSB [1971] 21:240–247) regards *Micromonospora fusca* Jensen 1932 as a subjective synonym of *Micromonospora purpureochromogenes* (Waksman and Curtis 1916) Luedemann 1971. These two species have different types.

Rule 24b

If two names compete for priority and if both names date from 1 January 1980 on an approved list, the priority shall be determined by the date of the original publication of the name before 1 January 1980.

Rule 24c

The Judicial Commission may place on the list of **rejected names** (*nomina rejicienda*) a name previously published in an approved list.

Rule 25a *Effective publication*

Effective publication is effected under this Code by making generally available, by sale or distribution, to the scientific community, printed material for the purpose of providing a permanent record.

Recommendation 25a

When a name of a new taxon is published in a work written in a language unfamiliar to the majority of workers in bacteriology, it is recommended that the author(s) include in the publication a description in a more familiar language.

Rule 25b

No other kind of publication than that cited in Rule 25a is accepted as effective nor are the following.

(1) Communication of new names at a meeting, in minutes of a meeting, or, after 1950, in abstracts of papers presented at meetings.

(2) Placing of names on specimens in collections or in listings or catalogues of collections.

(3) Distribution of microfilm, microcards, or matter reproduced by similar methods.

(4) Reports in ephemeral publications, newsletters, newspapers after 1900, or nonscientific periodicals.

(5) Inclusion of a name of a new taxon of bacteria in a published patent application or issued patent.

Rule 26a *Date of publication*

The date of publication of a scientific work is the date of publication of the printed matter. The date given to the work containing the name or epithet must be regarded as correct in the absence of proof to the contrary.

Rule 26b

The date of acceptance of an article for publication if given in a publication does not indicate the effective date of publication and has no significance in the determination of the priority of publication of names.

Valid and Invalid Publication

Rule 27

From the date of publication of this revision of the Code, a name of a new taxon, or a **new combination** for an existing taxon, is not validly published unless the following criteria are met.

(1) The name is published in the IJSB.

(2) The publication of the name in the IJSB is accompanied by a description of the taxon or by a reference to a previous effectively published description of the taxon (see Rules 25a and 25b, and for genus and species, Rules 29–32).

(3) The type is designated for a new taxon or cited for a new combination in the IJSB.

Note. After publication of this Code, valid publication of the name of a taxon requires publication in the IJSB of the name of the taxon and reference to an effectively published description whether in the IJSB or in another publication. The date of publication is that of publication in the IJSB. The name may be mentioned in a previous effectively published description, but the name is not validly published until its publication in the IJSB.

In the case of a name of a new taxon (rather than a new combination for a taxon already described), a type must be designated in the publication. It is recommended that the type of a species or subspecies be deposited in a recognized culture collection (see Recommendation 30a) and that the description of the taxon conform to minimal standards (see Recommendation 30b).

Rule 28a

An author validly publishing a new name after 1 January 1980 may revive a name validly published prior to 1 January 1980 (see Rule 24a) but not listed in one of the Approved Lists of Bacterial Names. The name may be used whether or not the new taxon is related in any way to the taxon to which the name was originally applied.

Authority for the name must be claimed by the new author. However, if the author wishes to indicate that the name is a revived name and is used to describe a taxon with the same circumscription, position, and rank as that given by the original author, he may do so by appending the abbreviation "nom. rev." (**revived name**) to the name (see Provisional Rule B3, p. 35).

Note 1. After 1 January 1980, publication of a new name is not invalidated by previous publication of the name (before 1 January 1980) unless the name is included in the Approved Lists of Bacterial Names.

Note 2. Since revived names are treated as new names, they require publication in the IJSB, and the date of valid publication of a revived name is that of the publication in the IJSB (see Rule 27).

Note 3. After 1 January 1980, search for publication of names and effectively published descriptions prior to 1 January 1980 will no longer be required. The Approved Lists of Bacterial Names will form the foundation of a new bacterial nomenclature and taxonomy.

Rule 28b

A name or epithet is not validly published which is:

(1) Not accepted at the time of publication by the author who published it.

Example: *Muellerina* de Petschenko 1910 (Opinion 10).

From the date of publication of this Code, names or epithets published with a question mark or other indication of taxonomic doubt yet accepted by the author are not validly published.

(2) Merely proposed in anticipation of the future acceptance of the taxon concerned or the acceptance of a particular circumscription, position, or rank for the taxon which is being named or in anticipation of the future discovery of some hypothetical taxon.

Examples: (a) *Clostrinium* Fischer 1895 (Opinion 20); (b) *Corynebacterium hemophilum* Svendsen *et al.* (J. Bacteriol. [1947] **53**:758). "Its haemophilic properties might be used in coining a name, and the name *Corynebacterium hemophilum* is suggested in case further investigation should justify its rank as a species."

(3) Mentioned incidentally. **Incidental mention** of a new name means mention by an author who does not clearly state or indicate that he is proposing a new name or combination.

Examples: (a) *Pseudobactrinium* Trevisan 1888. (b) Raj (IJSB [1970] **20**:79) stated: "Also, recently another organism tentatively named as *Microcyclus marinus* was isolated from the ocean."

Valid Publication of the Name of a Genus or Subgenus, including a Monotypic Genus

Rule 29

For a generic or subgeneric name to be validly published it must comply with the following conditions.

(1) It must be published in conformity with Rules 27 and 28b.

(2) The genus or subgenus named must include one or more described or previously described species.

Instead of a description of the genus or subgenus, a citation to a previously and effectively published description of the genus as a subgenus (or subgenus as a genus) may be given.

Example: Not yet found.

In the case of a genus containing a single species, a combined generic and specific description may be given.

Example: *Thiobacillus thioparus* Beijerinck 1904.

Recommendation 29

A description of a genus or subgenus should mention the points in which the genus or subgenus differs from related genera or subgenera. Where possible, the family to which it belongs should be mentioned.

Valid Publication of the Name of a Species

Rule 30

For the name of a species to be validly published, it must conform with the following conditions.

(1) It must be published in conformity with Rules 27 and 28b.

(2) It must be published as a binary combination consisting of a generic name followed by a single specific epithet (see Rule 12a).

Recommendation 30a

Before publication of the name of a new species, a culture of the type strain (or, if the species is noncultivable, type material, a photograph or illustration) should be deposited in at least one of the permanently established culture collections from which it would be readily available. The designation allotted to the strain by the culture collection should be quoted in the published description.

Recommendation 30b

Before publication of the name and description of a new species, the examination and description should conform at least to the **minimal standards** (if available) required for the relevant taxon of bacteria.

Note. Lists of **minimal standards** will be prepared for each group of bacteria by experts at the request of the Judicial Commission for consideration by the Judicial Commission and the ICSB for publication in the IJSB. Such standards will include tests for the establishment of generic identity and for the **diagnosis** of the species, i.e., an indication of characters which would distinguish the species from others.

Rule 31a

The name of a species or subspecies is not validly published if the description is based upon studies of a mixed culture of more than one species or subspecies. This does not apply to descriptions based chiefly on morphology (e.g., *Achromatium oxaliferum* Schewiakoff 1893).

Rule 31b

The name of a **consortium** is not regulated by this Code, and such a name has no standing in nomenclature.

Example: *Cylindrogloea bacterifera* Perfiliev 1914.

Note. A **consortium** is an aggregate or association of two or more organisms.

Valid Publication of the Name of a Subspecies

Rule 32a

For the name of a subspecies to be validly published, it must conform with the following conditions.

(1) It must be published in conformity with Rules 27 and 28b.

(2) It must be published as a **ternary combination** consisting of the generic name followed by a single specific epithet and this in turn by a single subspecific epithet, with the abbreviation "subsp." between the two epithets to indicate the rank (see Rule 13a).

Example: *Bacillus subtilis* subsp. *subtilis.*

(3) The author must clearly indicate that a subspecies is being named.

Recommendation 32a

Recommendations 30a and 30b apply to the name of a subspecies with replacement of the word "species" by the word "subspecies."

Publication of a Specific or Subspecific Epithet

Rule 32b

A specific (or subspecific) epithet is not rendered illegitimate by publication in a species (or subspecies) name in which the generic name is illegitimate (see also Chapter 3, Section 8 and example to Rule 20f).

Section 6. Citation of Authors and Names

Proposal and Subsequent Citation of the Name of a New Taxon

Rule 33a

An author should indicate that a name is being proposed for a new taxon by the addition of the appropriate abbreviation for the category to which the taxon belongs.

Note 1. Appropriate abbreviations are: **"ord. nov."** for *ordo novus,* **"gen. nov."** for *genus novum,* **"sp. nov."** for *species nova,* **"comb. nov."** for *combinatio nova.* Similar abbreviations may be formed as required.

Note 2. Although words or abbreviations in Latin are usually printed in italics, such abbreviations as the above are frequently printed in Roman or boldface type when they follow a Latin scientific name in order to differentiate them from the name and draw attention to the abbreviation.

Examples: Order, *Actinomycetales* ord. nov.; family, *Actinomycetaceae* fam. nov.; genus, *Actinomyces* gen. nov.; species, *Actinomyes bovis* sp. nov.

Rule 33b

The citation of the name of a taxon that has been previously proposed should include both the name of the author(s) who first published the name and the year of publication.

Example: *Actinomyces bovis* Harz 1877.

Note 1. Correct citation of a name enables the date of publication to be verified, the original description to be found, and the use of the name by different authors for different organisms to be distinguished.

Example: *Mycobacterium terrae* Wayne 1966, not *Mycobacterium terrae* Tsukamura 1966.

Note 2. Full citation of the publication should include reference to the page number(s) in the main text of the scientific work in which the name was proposed, not to the summary or abstract of that text even if proposal of the name is mentioned in that summary or abstract.

Example: *Bacillus subtilis* (Ehrenberg 1835) Cohn 1872, 174. The page number "174" is the page in Cohn's publication (Untersuchungen über Bacterien. Beitr. Biol. Pfl. Heft 2. 1:127–224) on which the proposal of the new combination occurs.

Proposal and Subsequent Citation of a New Combination

Rule 34a

When an author transfers a species to another genus (Rule 41), or a subspecies to another species, then the author who makes the transfer should indicate the formation of the **new combination** by the addition to the citation of the abbreviation **"comb. nov."** *(combinatio nova).*

This form of citation should be used when the author retains the original specific epithet in the new combination; however, if an author is obliged to substitute a new specific epithet as a result of homonymy, the abbreviation **"nom. nov."** (*nomen novum*) should be used [see Rule 41a(1)]. The original name is referred to as the **basonym.**

Example: *Actinomyces exfoliatus* Waksman and Curtis 1916; *Streptomyces exfoliatus* (Waksman and Curtis 1916) comb. nov. (It was correctly cited this way by Waksman and Henrici in *Bergey's Manual of Determinative Bacteriology*, 6th ed. The Williams & Wilkins Co., Baltimore, 1948.)

Rule 34b

The citation of a **new combination** which has been previously proposed should include the name of the original author in parentheses followed by the name of the author(s) who proposed the new combination and the year of publication of the new combination.

Example: *Bacillus polymyxa* (Prazmowski) Macé 1889 or *Bacillus polymyxa* (Prazmowski 1880) Macé 1889.

Note 1. The inclusion of the date of the publication of the original author of the name is to be preferred, although it is sometimes omitted since the date can be expected to be found in the publication of the author(s) who proposed the new combination.

Example: *Bacillus polymyxa* (Prazmowski 1880) Macé 1889 is to be preferred to *Bacillus polymyxa* (Prazmowski) Macé 1889.

Note 2. When, however, the author who formed the new combination was obliged to substitute a new specific epithet to avoid homonymy [see Rule 41a(1)], the name of the author of the original specific epithet is omitted.

Example: *Streptomyces aurioscleroticus* Pridham 1970 **not** *Streptomyces aurioscleroticus* (Thirumalachar *et al.* 1966) Pridham 1966 [see Example to Rule 41a(1) for explanation].

Rule 34c

When a taxon from subspecies to genus is altered in rank but retains its name or epithet, the original author(s) must be cited in parentheses followed by the name of the author(s) who effected the alteration and the year of publication.

Example: *Actinomyces exfoliatus* Waksman and Curtis 1916 to *Actinomyces chromogenes* subsp. *exfoliatus* (Waksman and Curtis 1916) Krasil'nikov 1941.

Citation of the Name of a Taxon whose Circumscription has been Emended

Rule 35

If an alteration of the diagnostic characters or of the circumscription of a taxon modifies the nature of the taxon, the author responsible may be indicated by the addition to the author citation of the abbreviation **"emend."** (*emendavit*) followed by the name of the author responsible for the change.

Example: *Rhodopseudomonas* Czurda and Maresch 1937 emend. van Niel 1944 (see Opinion 49).

Citation of a Name conserved so as to exclude the Type

Rule 36

A name conserved so as to exclude the type is not to be ascribed to the original author, but the author whose concept of the name is conserved must be cited as authority.

Example: *Aeromonas liquefaciens*, the type species of the genus *Aeromonas*, has been excluded from *Aeromonas* (Opinion 48). The generic name *Aeromonas* is now attributed to Stanier 1943, not to Kluyver and van Niel 1936.

Citation of Names included in the Approved Lists of Bacterial Names, 1 January 1980

(As no Approved Lists of Bacterial Names exist, the following rules can only be provisional.)

Note. The nomenclatural status of a name included in an approved list is the same as that of new names published after 1 January 1980. If special reference to the inclusion of a name in an approved list be desired, Provisional Rules A1–2 and B1–4 should be followed (p. 33–35).

Provisional Rule A1

(1) The citation of a name which is included in an approved list can include the name of the original author and date of publication followed by the words "Approved List No. . . . 1980" in parentheses.

Examples: *Bacillus cereus* Frankland & Frankland 1888 (Approved List No. 1, 1980); *Bacillus subtilis* (Ehrenberg 1835) Cohn 1872 (Approved List No. 1, 1980).

(2) Alternatively, a name which is included in an approved list may

be cited simply by the addition of the words "Approved List No. . . . , 1980" in parentheses.

Examples: *Bacillus cereus* (Approved List No. 1, 1980); *Bacillus subtilis* (Approved List No. 1, 1980).

(3) If indication is given that a name is included in an approved list without specification of that list, the abbreviation **"nom. approb."** (*nomen approbatum*) may be appended to the name of the taxon.

Example: *Bacillus subtilis* nom. approb.

Provisional Rule A2

If an author transfers a species which has been included in one of the approved lists to another genus, the proposal of the **new combination** should be made by the addition of the abbreviation **"comb. nov."** (*combinatio nova*) followed in parentheses by the name under which it appeared in the approved list and the correct citation for that list.

Example: If *Bordetella parapertussis* appears in Approved List No. 5, 1980 and is transferred by Smith in 1983 to the genus *Moraxella*, the citation by Smith may be as follows: *Moraxella parapertussis* (Eldering and Kendrick 1938) comb. nov. (*Bordetella parapertussis*—Approved List No. 5, 1980). Another author citing this proposal would then use the citation: *Moraxella parapertussis* (Eldering and Kendrick 1938) Smith 1983 (*Bordetella parapertussis*—Approved List No. 5, 1980).

Note. The correct form of citation of an approved list may be found in the approved list itself.

Citation of Reused Names

(Since the reuse of names has not come into operation, the following rules can only be provisional.)

Provisional Rule B1

If a name or epithet which was published prior to 1 January 1980 but not included in an approved list is proposed by an author for a different or for the same taxon, the name or epithet must be attributed to the author of the proposal (Rule 28a), and the citation should be made according to Rules 33a, b and 34a, b.

Provisional Rule B2

If a name or epithet is revived for the same taxon (in the author's opinion), the author may indicate the fact by addition of the abbreviation

"nom. rev." (*nomen revictum*) after the correct abbreviation (Rule 33a) for the category concerned.

Example: *Bacillus palustris* sp. nov. nom. rev.

Provisional Rule B3

If an author wishes to indicate the names of the original authors of a revived name, he may do so by citation of the name of the taxon, followed by the word **"ex"** and the name of the original author and the year of publication, in parentheses, followed by the abbreviation "nom. rev."

Example: *Bacillus palustris* (*ex* Sickles and Shaw 1934) nom. rev. A subsequent author citing this revived name would use the citation *Bacillus palustris* Brown 1982, or *Bacillus palustris* (*ex* Sickles and Shaw 1934) Brown 1982.

Provisional Rule B4

If an author wishes to indicate that a reused name has been used for a different taxon, indication is made by citation of the name and the author and year of publication followed by the word **"non"** (or "not") and the name and year of publication of the author who first used the name.

Section 7. Changes in Names of Taxa as a Result of Transference, Union, or Change in Rank

Rule 37a

(1) The name of a taxon must be changed if the nomenclatural type of the taxon is excluded.

(2) Retention of a name in a sense which excludes the type can only be effected by conservation and only by the Judicial Commission (see also Rule 23a). At the time of conservation, the new type is established by the Judicial Commission.

Rule 37b

A change in the name of a taxon is not warranted by an alteration of the diagnostic characters or of the circumscription. A change in its name may be required by one of the following.

(1) An Opinion of the Judicial Commission (see Rule 37a(1) above).
(2) Transfer of the taxon (see Rule 41).
(3) Union with another taxon (Rules 42-44, 47a, and 48b).
(4) Change of its rank (Rules 48, 49, 50a, 50b).

Rule 38

When two or more taxa of the same rank are united, then the name of the taxon under which they are united (and therefore the type of the taxon) is chosen by the rule of priority of publication.

Example: White 1930 united *Eberthella* Bergey *et al.* 1923 with *Salmonella* Lignières 1900 and retained the earlier name, *Salmonella.*

Note. Eberthella was raised by Bergey *et al.* 1923 to a genus from the subgeneric name, *Eberthella* Buchanan 1918.

If, however, this choice would lead to confusion in bacteriology, the author should refer the matter to the Judicial Commission. (For taxa above the rank of species, see also Rule 47a.)

Example: Not yet found.

Division of a Genus into Genera or Subgenera, and of a Subgenus into Subgenera

Rule 39a

If a genus is divided into two or more genera or subgenera, the generic name must be retained for one of these. If the name has not been retained (in a previous publication), it must be reestablished under Rule 39b. (See Rule 49 when a subgenus is raised to genus.)

Example: If the genus *Bacillus* is divided into the two subgenera *Bacillus* and *Aerobacillus,* the subgenus which includes the type species *Bacillus subtilis* must be named *Bacillus.*

Rule 39b

When a particular species has been designated as the type, the generic name must be retained for the genus which includes that species. When no type was designated a type must be chosen.

Rule 39c

The principles of Rules 39a and 39b apply when a subgenus is divided into two or more subgenera, the original subgeneric name being retained for that subgenus which contains the type species.

Division of a Species into Species or Subspecies, and of a Subspecies into Subspecies

Rule 40a

When a species is divided into two or more species or subspecies, the specific epithet of the original species must be retained for one of the taxa into which the species is divided or, if the epithet has not been retained (in a previous publication), it must be reestablished. (See Rule 50a when a subspecies is elevated to a species.)

Rule 40b

The specific epithet must be retained for the species or subspecies which includes the type strain. When no type was designated, one must be chosen.

Example: If the species *Bacillus subtilis* is divided into subspecies, the subspecies containing the type strain must be named *Bacillus subtilis* subsp. *subtilis*.

Rule 40c

The principles of Rules 40a and 40b apply when a subspecies is divided into two or more subspecies, the original subspecific name being retained for that subspecies which contains the type strain.

Note. Although the specific and subspecific epithets in the name of a type subspecies are the same, they do not contravene Rule 12b because they are based on the same type.

Transfer of a Species to another Genus

Rule 41a

When a species is transferred to another genus without any change of rank, the specific epithet must be retained, or if it has not been retained (in a previous publication), it must be reestablished, unless:

(1) The resulting binary combination would be a **junior homonym.**

Example: Pridham (1970) proposed *Streptomyces aurioscleroticus* for *Chainia aurea* Thirumalachar *et al.* 1966 on transfer to *Streptomyces* because in that genus the name *Streptomyces aureus* already existed.

(2) There is available an earlier validly published and legitimate specific or subspecific epithet.

Example: Not yet found.

Rule 41b

When the name of a genus is changed, the specific epithets of the species included under the original generic name must be retained for the same species if they are transferred to the new genus.

Union of Taxa of Equal Rank

Rule 42

In the case of subspecies, species, subgenera and genera, if two or more of those taxa of the same rank are united, the oldest legitimate name or epithet is retained.

If the names or epithets are of the same date, the author who first unites the taxa has the right to choose one of them, and his choice must be followed.

Recommendation 42

Authors who have to choose between two generic names of the same date should note the following.

(1) Prefer the one which is better known.

(2) Prefer the one which was first accompanied by the description of a species.

(3) If both are accompanied by descriptions of species, prefer the one which includes the larger number of species.

(4) In cases of equality from these points of view, prefer the more appropriate name.

Union of Genera as Subgenera

Rule 43

When several genera are united as subgenera of one genus, the subgenus which includes the type species of the genus under which union takes place must bear the same name as that genus.

Example: The subgenus name *Lactobacillus* Beijerinck 1901 must be used instead of *Thermobacterium* for the subgenus that contains the type species *Lactobacillus delbrueckii* (see *Bergey's Manual*, 1957, p. 543 and Opinion 38).

Union of Species of Two or More Genera as a Single Genus

Rule 44

If two or more species of different genera are brought together to form a genus, and if these species include the type species of one or more genera, the name of the genus is that associated with the type species having the earliest legitimate generic name.

If no type species is placed in the genus, a new generic name must be proposed and a type species selected.

Example: *Brevibacterium* Breed 1953. None of the included species was a type species of the genera from which the species were transferred, so a new name *Brevibacterium* was proposed, with *Brevibacterium linens* as the type species.

Union of Species as Subspecies

Rule 45

When several species are united as subspecies under one species, the subspecies which includes the type strain of the species under whose name they are united must be designated by the same epithet as the species.

Example: *Streptomyces griseus* subsp. *griseus* [see Int. Bull. Bacteriol. Nomencl. Taxon. (1965) **15**:214 and 224].

Rule 46

The valid publication of a subspecific name which excludes the type of the species automatically creates another subspecies which includes the type and whose name bears the same specific and subspecific epithets as the name of the type.

Example: Publication of *Bacillus subtilis* subsp. *viscosus* Chester 1904 automatically created a new subspecies, *Bacillus subtilis* subsp. *subtilis*.

The author of the species name is to be cited as the author of such an automatically created subspecific name.

Example: *Bacillus subtilis* subsp. *subtilis* (Ehrenberg 1835) Cohn 1872.

Union of Taxa above Species under a Higher Taxon

Rule 47a

When two or more taxa of the same rank from subtribe to family inclusive are united under a taxon of higher rank, the higher ranking taxon should derive its name from the name of the earliest legitimate genus that is a type genus of one of the lower ranking taxa.

If, however, the use of this generic name would lead to confusion in bacteriology, then the author may choose as type a genus which, in his opinion, leads to the least confusion and, if in doubt, should refer the matter to the Judicial Commission.

Note. The type of a taxon above the rank of genus is one of the contained genera (Rule 15). The name of the type subgenus is the same as that of the type genus; therefore, only the names of genera need to be considered.

Example: Buchanan in Breed *et al.* 1957 followed the law of priority in combining the families *Beggiatoaceae* Migula 1894 and *Vitreoscillaceae* Pringsheim 1949 into the new order *Beggiatoales,* whose type is *Beggiatoa* Trevisan 1842, which has priority over *Vitreoscilla* Pringsheim 1949. In contrast, Breed *et al.* 1957 chose *Pseudomonas* Migula 1894 over *Spirillum* Ehrenberg 1832 and *Nitrobacter* Winogradsky 1892 to form the name of a new suborder, *Pseudomonadineae* Breed *et al.* 1957.

Rule 47b

If no type genera were placed in the taxon, a new name based on the selected type must be proposed for the taxon.

Example: *Peptococcaceae* Rogosa 1971 [see IJSB (1971) **21**:235].

Change in Rank

Rule 48

When the rank of a taxon between subgenus and order is changed, the stem of the name must be retained and only the suffix altered unless the resulting name must be rejected under the Rules (see Rule 9).

Example: Elevation of the tribe *Pseudomonadeae* to the family *Pseudomonadaceae.*

Rule 49

When a genus is lowered in rank to subgenus, the original name must be retained unless it is rejected under the Rules. This also applies when a subgenus is elevated to a genus.

Example: If the genus *Aerobacillus* is lowered in rank to subgenus, the name of the subgenus is *Aerobacillus.*

Rule 50a

When a subspecies is elevated in rank to a species, the subspecific epithet in the name of the subspecies must be used as the specific epithet of the name of the species unless the resulting combination is illegitimate.

Example: *Propionibacterium jensenii* subsp. *raffinosaceum* van Niel 1928 becomes *Propionibacterium raffinosaceum* (van Niel) Werkman and Kendall 1931.

Rule 50b

When a species is lowered in rank to a subspecies, the specific epithet in the name of the species must be used as the subspecific epithet of the name of the subspecies unless the resulting combination is illegitimate.

Example: *Bacillus aterrimus* Lehmann and Neumann 1896 becomes *Bacillus subtilis* subsp. *aterrimus* (Lehmann and Neumann 1896) Smith *et al.* 1946.

Section 8. Illegitimate Names and Epithets; Replacement, Rejection, and Conservation of Names and Epithets

Illegitimate Names

Rule 51a

A name contrary to a Rule is illegitimate and may not be used. However, a name of a taxon which is illegitimate when the taxon is in one taxonomic position is not necessarily illegitimate when the taxon is in another taxonomic position.

Example: If the genus *Diplococcus* Weichselbaum 1886 is combined with the genus *Streptococcus* Rosenbach 1884, *Diplococcus* is illegitimate as the name of the combined genus because it is not the earlier name. If the genus *Diplococcus* Weichselbaum is accepted as separate and distinct, then the name *Diplococcus* is legitimate.

Rule 51b

Among the reasons for which a name may be illegitimate are the following.

(1) If the taxon to which the name was applied, as circumscribed by the author, included the nomenclatural type of a name which the author ought to have adopted under one or more of the Rules.

Example: If an author circumscribes a genus to include *Bacillus subtilis,* the type species of the genus *Bacillus,* then the circumscribed genus must be named *Bacillus.*

(2) If the author did not adopt for a binary or ternary combination the earliest legitimate generic name, specific epithet, or subspecific epithet available for the taxon with its particular **circumscription, position,** and **rank.**

Example: The name *Bacillus whitmori* Bergey *et al.* 1930 was illegitimate as Whitmore had named the organism *Bacillus pseudomallei* in 1913.

(3) If its specific epithet must be rejected under Rules 52 or 53.

(4) If it is a **junior homonym** of a name of a taxon of bacteria, fungi, algae, protozoa, or viruses.

Example: *Phytomonas* Donovan 1909, a genus of flagellates, antedates *Phytomonas* Bergey *et al.* 1923, a genus of bacteria (Opinion 14).

Note. After 1 January 1980, names of bacteria validly published under this revision of the Code are not to be rejected as **homonyms** of names of bacteria published under previous editions of the Bacteriological Code.

Illegitimate Epithets

Rule 52

The following are not to be regarded as specific or subspecific epithets.

(1) A word or phrase which is not intended as a specific epithet.

Example: *Bacillus nova species* Matzuschita.

(2) A word which is merely an ordinal adjective used for enumeration.

Example: *primus, secundus.*

(3) A number or letter.

Example: α in *Bacillus* α von Freudenreich.

Rule 53

An epithet is illegitimate if it duplicates a specific or subspecific epithet previously validly published for a species or subspecies of the same genus but which is a different bacterium whose name is based upon another type.

Example: *Corynebacterium helvolum* (Zimmermann 1890) Kisskalt and Berend 1918 is based on the type of *Bacillus helvolus* Zimmermann 1890; the specific epithet *helvolum* cannot be used for *Corynebacterium helvolum* Jensen 1934, which is a different bacterium whose name is based upon another type.

Replacement of Names

Rule 54

A name or epithet illegitimate according to Rules 51b, 53, or 56a is replaced by the oldest legitimate name or epithet in a **binary** or **ternary combination** which in the new position will be in accordance with the Rules.

If no legitimate name or epithet exists, one must be chosen. Since a specific epithet is not rendered illegitimate by publication in a species name in which the generic name is illegitimate (Rule 32b), an author may use such an epithet if he wishes, provided that there is no obstacle to its employment in the new position or sense; the resultant combination is treated as a new name and is to be ascribed to the author of the combination. The epithet is, however, ascribed to the original author.

Example: *Pfeifferella pseudomallei* (Whitmore 1913) Ford 1928 is an illegitimate combination since *Pfeifferella* is a homonym of a protozoan generic name (Opinion 14). The epithet *pseudomallei* can be used for this organism in another genus, *Pseudomonas pseudomallei* (Whitmore 1913) Haynes 1957.

Rule 55

A legitimate name or epithet may not be replaced merely because of the following.

(1) It is inappropriate.
Example: *Bacteroides melaninogenicum* (see J. Gen Microbiol. **1**:109–120 [1947]).

(2) It is disagreeable.
Example: *Miyagawanella lymphogranulomatis.*

(3) Another name is preferable.
Example: Not yet found.

(4) Another name is better known.
Example: *Corynebacterium pseudodiphtheriticum* cannot be rejected because the synonym *Corynebacterium hofmanni* is better known.

(5) It no longer describes the organism.
Example: *Haemophilus influenzae.*

(6) It has been cited incorrectly: an incorrect citation can be rectified by a later author.
Example: *Proteus morgani* Yale 1939 (see Lessel, E. F., IJSB **21**:55–57 [1971]).

Rejection of Names

Rule 56a

Only the Judicial Commission can place names on the list of **rejected names** (*nomina rejicienda*) (see Rule 23a, Note 4 and Appendix 4). A name may be placed on this list for various reasons, including the following.

(1) An **ambiguous name** (*nomen ambiguum*), i.e., a name which has been used with different meanings and thus has become a source of error.

Example: *Aerobacter* Beijerinck 1900 (Opinion 46).

(2) A **doubtful name** (*nomen dubium*), i.e., a name whose application is uncertain.

Example: *Leuconostoc citrovorum* (Opinion 45).

(3) A **name causing confusion** (*nomen confusum*), i.e., a name based upon a mixed culture.

Example: *Malleomyces* Hallier 1870.

(4) A **perplexing name** (*nomen perplexum*), a name whose application is known but which causes uncertainty in bacteriology (see Rule 57c).

Example: *Bacillus limnophilus* Bredemann and Stürck *in* Stürck 1935 (Greek-Greek, marsh loving) and *Bacillus limophilus* Migula 1900 (Latin-Greek, mud-loving); see *Index Bergeyana*, p. 196.

Conservation of Names

Rule 56b

A **conserved name** (*nomen conservandum*) is a name which must be used instead of all earlier synonyms and homonyms.

Note 1. A conserved name (*nomen conservandum*) is conserved against all other names for the taxon, whether these are cited in the corresponding list of rejected names or not, so long as the taxon concerned is not united with another taxon bearing a legitimate name. In the event of union or reunion with another taxon, the earlier of the two competing names is adopted in accordance with Rules 23a, b.

Note 2. Only the Judicial Commission can place names on the list of **conserved names** (*nomina conservanda*) (see also Rule 23a, Note 4 and Appendix 4).

Section 9. Orthography

Rule 57a

Any name or epithet should be written in conformity with the spelling of the word from which it is derived and in strict accordance with the rules of Latin and latinization. Exceptions are provided for typographic and orthographic errors in Rule 61 and orthographic variants in Rules 62a and 62b.

Rule 57b

In this Code, **orthographic variant** means a name (or epithet) which differs from another name only in transliteration into Latin of the same word from a language other than Latin or in its grammatical correctness.

Example: *Haemophilus, Hemophilus.*

Rule 57c

When two or more generic names, or two or more epithets in the same genus are so similar (although the words are from different sources) as to cause uncertainty, they may be treated as **perplexing names** (*nomina perplexa*) and the matter referred to the Judicial Commission (see Rule 56a [4]).

Example: *Bacillus limnophilus* and *Bacillus limophilus* (see Rule 56a [4]).

Note 1. **Orthographic variants** may be corrected by any author.

Note 2. **Perplexing names** may be placed on the list of rejected names only by the Judicial Commission, because decisions on the status of names derived from different sources differing in one or more letters affect many well-known names in bacteriology.

Example: *Salmonella gamaba* and *Salmonella gambaga.*

Rule 58

When there is doubt about different spellings of the same name or epithet, or where two spellings are sufficiently alike to be confused, the question should be referred to the Judicial Commission which may issue an **Opinion** as seems fit. If one of the spellings is preferred by the Judicial Commission, this spelling should be used by succeeding authors.

Example: The epithet "*megaterium*" (over "*megatherium*") in the species name *Bacillus megaterium* du Bary 1884 (Opinion 1).

Rule 59

An epithet, even one derived from the name of a person, should not be written with an initial capital letter.

Example: *Shigella flexneri* (named after Flexner).

Rule 60

Intentional latinizations involving changes in orthography of personal names, particularly those of earlier authors, must be preserved.

Example: Pasteur may be latinized as Pastor, and *Streptococcus pastorianus* is derived from Pastor.

Typographic and Orthographic Errors

Rule 61

The **original spelling** of a name or epithet must be retained, except typographic or orthographic errors. Original spelling does not refer to the use of an initial capital or small letter or to diacritic signs.

Example: The original spelling was *Bacillus megaterium,* not *megatherium* (Opinion 1).

An unintentional typographic or orthographic error later corrected by the author is to be accepted in its corrected form without affecting its validity and original date of publication. It can also be corrected by a subsequent author who may or may not mention that the spelling is corrected, but the abbreviation **"corrig."** (*corrigendum*) may be appended to the name if an author wishes to draw attention to the correction. Succeeding authors may be unaware that the original usage was incorrect and use the spelling of the original author(s). Other succeeding authors may follow the correction of a previous author or may independently correct the spelling themselves, but in no case is the use of corrig. regarded as obligatory. None of these corrections affects the validity and original date of publication.

Example: *Mycobacterium stercusis* (sic) Bergey *et al.* 1923. Typographic error later corrected by the authors to *Mycobacterium stercoris;* this may be cited as *Mycobacterium stercoris* corrig.

Note. The liberty of correcting a name or epithet under Rules 61, 62a, and 62b, must be used with reserve especially if the change affects the first syllable and above all the first letter of the name or epithet.

Orthographic Variants by Transliteration

Rule 62a

Words differing only in transliteration into Latin from other languages which do not use the Latin alphabet are to be treated as **orthographic variants** unless they are used as the names of taxa based upon different types, when they are to be treated as **homonyms.** For an account of possible orthographic variants, see Appendix 9.

Example: *Haemophilus* and *Hemophilus.*

Rule 62b

When there are orthographic variants based on the same type, and there is no clear indication that one is correct, then an author has the right of choice.

Personal Names

Rule 63

The genitive and adjectival forms of a personal name are treated as different epithets and not as orthographic variants unless they are so similar as to cause confusion. For the latinization of personal names, see Appendix 9.

Example: The epithets *pasteurii* (genitive noun) and *pasteurianum* (adjective) are treated as different epithets.

Diacritic Signs

Rule 64

Diacritic signs are not used in names or epithets in bacteriology.

In names or epithets derived from words with such signs, the signs must be suppressed and the letters transcribed as follows: (1) *ä, ö* and *ü* become *ae, oe* and *ue*; (2) *é, è* and *ê* become *e*; (3) *ø, æ* and *å* become *oe, ae* and *aa*, respectively.

Gender of Names

Rule 65

The gender of generic names is governed by the following.

(1) A Latin or Greek word adopted as a generic name retains the classical gender of its language of origin. Authors are recommended to give the gender of any proposed generic name.

Example: *Sarcina* (Latin feminine noun, a package).

In cases where the classical gender varies, the author has the right of choice between the alternatives (but see Opinion 3 for the masculine gender of *-bacter*).

Example: Not yet found.

Doubtful cases should be referred to the Judicial Commission.

Example: Not yet found.

(2) Generic or subgeneric names which are modern compounds from two or more Latin or Greek words take the gender of the last component of the compound word.

Example: *Lactobacillus* (masculine) milk rodlet from Latin: *lac, lactis* (neuter), milk; and *bacillus* (masculine), little staff.

If the ending is altered, the gender is that of the new ending in the language of origin.

Example: Not yet found.

(3) Arbitrarily formed generic names or vernacular names used as generic names take the gender assigned to them by their authors. When the original author failed to indicate the gender, a subsequent author has the right of choice.

Example: *Ricolesia* Rake 1957, who assigned the feminine gender.

Chapter 4

Advisory Notes

Chapter 4. ADVISORY NOTES

A. Suggestions for Authors and Publishers

Publishers of periodicals and books are requested to indicate the year, month, and day of publication either on the publication itself or, in the case of a periodical, on the succeeding number. This information, as well as the title of the periodical or book from which the paper is reproduced, should also be printed on separates, tear sheets, or reprints.

Separates or reprints should always bear the pagination of the periodical of which they form a part.

An author who describes and names a new taxon should indicate the rank of the taxon concerned and where possible the rank and name of the next higher taxon (e.g., the name of the family to which a new genus is allocated or the name of the order in which a new family is placed). The title of the work concerned should indicate that a new name is published even if the name itself is not quoted in the title.

Note. Valid publication of a new name or combination requires announcement in the IJSB (Rule 27).

It is important that descriptions and illustrations of new species be as complete as possible and conform to the minimal standards when available (see Recommendation 30b).

For scientific names of taxa, conventions shall be used which are appropriate to the language of the country and to the relevant journal and publishing house concerned. These should preferably indicate scientific names by a different type face, e.g., italic, or by some other device to distinguish them from the rest of the text.

The name of a genus should be spelled without abbreviation the first time it is used with a specific epithet in a publication and in the summary of that publication.

Example: *Bacillus subtilis.*

In a series of species names all belonging to the same genus, it is customary to abbreviate the name of the genus in all but the first species, even if it is the first mention of the succeeding species.

Example: *Bacillus subtilis, B. polymyxa.*

Later use of the name of the species previously cited usually has the name of the genus abbreviated, commonly to the first letter of the generic name.

Example: *B. subtilis.*

If, however, species are listed belonging to two or more genera which have the same initial letter, the generic names should be used in full.

B. Quotations of Authors and Names

(1) *Multiple authorship* **(et al.).** When the new name of a taxon is published under two authors, both are cited; when there are more than two authors and when there is no definite designation of a single individual as the author of the name, the citation may be made by listing the names of all the authors, or by giving the name of the first author, followed by the abbreviation "*et al.*" (*et alii*).

(2) *Publication in the work of another author* **(in).** When a new name or combination by one author is published in a work of another author, the word "*in*" should be used in the literature cited to connect the names of the two authors. The name of the author of the name of the taxon precedes the name of the author in whose work it is contained.

Example: *Streptomyces reticuli* Waksman, S. A., and A. T. Henrici *in* Breed, R.S., *et al. Bergey's Manual of Determinative Bacteriology*, 6th ed., 1948. The Williams & Wilkins Co., Baltimore.

(3) Use of **"pro synon.," "ex," "non,"** and **"sic."**

(a) When citing a name published as a synonym, the words "as synonym" or **"pro synon."** should be added to the citation. (For types of **synonym,** see Rule 24a.)

Example: *Pseudomonas pyocyanea* pro synon. *Pseudomonas aeruginosa.*

(b) When an author has published a name from a manuscript of another author, whether as a synonym or not, the word "*ex*" should be used to connect the names of the two authors. The name of the author who publishes the name precedes that of the original author.

Example: *Bacillus caryocyaneus* Dupaix 1930 *ex* Beijerinck (see Dupaix, 1930, Trav. Lab. Fac. Pharm., fasc. 3, p. 13).

(c) When citing in synonymy a name invalidated by an earlier homonym, the citation should be followed by the name of the author of the earlier homonym preceded by the word "*non,*" preferably with the date of publication added.

Example: *Pfeifferella* Buchanan 1918 *non* Labbé 1899.

(d) If a name or epithet is adopted with alterations from the form as originally published, including the use of a corrected spelling, the original spelling should be cited in any list of synonyms of the corrected name. The original spelling is followed by the term "*sic*" in parentheses to indicate that the original spelling is accurately cited.

Example: *Bacillus pantothenticus* (*sic*).

(4) *Nomen nudum.* In the citation of a **bare name** (*nomen nudum*), the status of the name should be indicated by adding "nom. nud."

Note. A **bare name** (*nomen nudum*) means a name published without a description or a reference to a previously published description.

Example: Not yet found.

(5) *Nomen conservandum.* A **conserved name** (*nomen conservandum*) shall be indicated by the addition of the abbreviation "nom. cons." to the citation.

Example: *Pseudomonas* Migula 1894 nom. cons. (Opinion 5).

C. Maintenance of Type Strains

The utmost importance should be given to the preservation of the original "type" material on which the description of a new species or subspecies is based.

Preserved and living specimens should be maintained in a bacteriological laboratory, more particularly in one of the permanently established culture collections, and a record of this fact should be included in the publication (see Recommendation 30a).

Maintenance may be by a variety of methods, e.g., in a medium, in a host by passage, in cells or exudates, or in the frozen or dried state.

Every precaution should be taken to maintain such cultures with a minimum amount of change. Repeated subculture may lead to phenotypic or genotypic changes.

Appendix 1

Codes of Nomenclature

Appendix 1. Codes of Nomenclature

International Code of Nomenclature of Bacteria and Viruses (1958) Ames, Iowa: Iowa State College Press. 186 pp.

Editorial Board. 1966. International Code of Nomenclature of Bacteria. Int. J. Syst. Bacteriol. **16**:459–490.

International Code of Botanical Nomenclature, adopted by the Eleventh International Botanical Congress, Seattle, August 1969. (International Bureau for Plant Taxonomy and Nomenclature, Utrecht, 1972, 426 pp.)

International Code of Nomenclature for Cultivated Plants. (International Bureau for Plant Taxonomy and Nomenclature, Utrecht, 1969, 32 pp.)

International Code of Zoological Nomenclature, adopted by the XV International Congress of Zoology. (International Commission on Zoological Nomenclature, London, 1964, 176 pp.)

The Code of Nomenclature of Viruses has not yet been issued, but see Classification and Nomenclature of Viruses, First Report of the International Committee on Nomenclature of Viruses, edited by P. Wildy. (S. Karger, Basel, 1971, 81 pp.)

Appendix 2

Approved Lists of Bacterial Names

Appendix 2. Approved Lists of Bacterial Names

References to the Approved Lists of Bacterial Names in the IJSB may be added to this Appendix in future editions of the International Code of Nomenclature of Bacteria.

Appendix 3

Published Sources for Names of Bacterial, Algal, Protozoal, Fungal, and Viral Taxa

Appendix 3. Published Sources for Names of Bacterial, Algal, Protozoal, Fungal, and Viral Taxa

The following publications are among the major sources for names of bacterial, algal, protozoal, fungal, and viral taxa.

Bacteria

Buchanan, R. E., J. G. Holt, and E. F. Lessel. 1966. *Index Bergeyana* (The Williams & Wilkins Co., Baltimore, 1472 pp.)

Hatt, H. D., and E. Zvirbulis. 1967. Status of names of bacterial taxa not evaluated in *Index Bergeyana* (1966). I. Names published *circa* 1950–1967 exclusive of the genus *Salmonella.* Int. J. Syst. Bacteriol. **17**:171–225.

Zvirbulis, E., and H. D. Hatt. 1969. Status of names of bacterial taxa not evaluated in *Index Bergeyana* (1966). Addendum II. *Acetobacter* to *Butyrivibrio.* Int. J. Syst. Bacteriol. **19**:57–115.

Zvirbulis, E., and H. D. Hatt. 1969. Status of names not evaluated in *Index Bergeyana* (1966). Addendum III. *Achromobacter* to *Lactobacterium.* Int. J. Syst. Bacteriol. **19**:309–369.

Algae

De Toni, 1889. Sylloge Algarum.

Index Kewensis. 1895-present. (Royal Botanical Gardens, Kew.)

Protozoa

Nomenclator Zoologicus. 1758–present. Published in four volumes and two supplements from 1939 onwards. Edited by S. A. Neave. Zoological Society, London.

Index Zoologicus. 1800–1900. Charles Owen Waterhouse. (Published 1902). Edited by David Sharpe. Zoological Society, London.

Index Zoologicus. 1902-present. (Zoological Society, London.)

Fungi

Saccardo, P. A. 1882–1921. Sylloge Fungorum. (Pavia, 25 vol.)

Clements, F. E., and C. L. Shear. 1931. The Genera of Fungi (H. W. Wilson & Co., New York.)

Index to Fungi. 1940-present. (Commonwealth Mycological Institute, Kew.)

Petrak's Lists of Fungal Names. 1922–1940. Available from Commonwealth Mycological Institute, Kew, Surrey, England.

Ainsworth, G. C. 1971. Ainsworth and Bisby's Dictionary of Genera of the Fungi. Commonwealth Mycological Institute, Kew, Surrey, England.

Viruses

Wildy, P. (editor) 1971. Classification and Nomenclature of Viruses. First Report of the International Committee on Nomenclature of Viruses. (S. Karger, Basel, 81 pp.)

Appendix 4

Conserved and Rejected Names of Bacterial Taxa

Appendix 4. Conserved and Rejected Names of Bacterial Taxa

(Nomina taxorum conservanda et rejicienda)

LIST 1. *Conserved and rejected family names.*
LIST 2. *Conserved names of genera of bacteria.*
LIST 3. *Conserved specific epithets in names of species of bacteria.*
LIST 4. *Rejected names of genera and subgenera of bacteria.*
LIST 5. *Rejected specific epithets in names of species of bacteria.*

The citations are (unless otherwise indicated) to the volumes, pages, and dates of the *International Bulletin of Bacteriological Nomenclature and Taxonomy* up to vol. 15 (1965) and thereafter of the *International Journal of Systematic Bacteriology.*

LIST 1. *Conserved and rejected family names (nomina familiarum conservanda et rejicienda)*

Conserved name (*nomen conservandum*)	Name of type genus of conserved family	Rejected name (*nomen rejiciendum*)	Opinion no.	Citation
Enterobacteriaceae	*Escherichia* Castellani and Chalmers 1919, p. 841	*Bacteriaceae*	15	**8**:73–74 (1958)

LIST 2. *Conserved names of genera of bacteria (nomina generum bacteriorum conservanda)*

Conserved generic names (*nomina generum conservanda*)	Name of type species of conserved genus	Opinion no.	Citations
Aeromonas Stanier 1943	*Aeromonas hydrophila* (Chester) Stanier 1943	48	**23**:473–474 (1973)
Agrobacterium Conn 1942	*Agrobacterium tumefaciens* (Smith and Townsend) Conn 1942	33	**20**:10 (1970)
Arthrobacter Conn and Dimmick 1947	*Arthrobacter globiforme* (Conn) Conn and Dimmick 1947	24	**8**:171–172 (1958)

LIST 2. *Conserved names of genera of bacteria (nomina generum bacteriorum conservanda)* (continued)

Conserved generic names (*nomina generum conservanda*)	Name of type species of conserved genus	Opinion no.	Citations
Bacillus Cohn 1872	*Bacillus subtilis* Cohn 1872	A. (1936)	Proc. 2nd. Internatl. Congr. Microbiol. London, 1936; Journal of Bacteriology, **33**:445 (1937), International Code of Nomenclature of Bacteria and Viruses (1958) p. 148
Beggiatoa Trevisan 1842	*Beggiatoa alba* (Vaucher) Trevisan 1845, *Oscillatoria alba* Vaucher 1803	13	4:151–156 (1954)
Chlorobacterium Lauterborn 1916	*Chlorobacterium symbioticum* Lauterborn 1916	6	4:143 (1954)
Chromobacterium Bergonzini 1880	*Chromobacterium violaceum* Bergonzini 1880	16	**8**:151–152 (1958)
Enterobacter Hormaeche and Edwards 1960	*Enterobacter cloacae* (Jordan) Hormaeche and Edwards 1960	28	**13**:38 (1963)
Escherichia Castellani and Chalmers 1919	*Escherichia coli* Castellani and Chalmers 1919 (**basonym** *Bacillus coli* Migula 1895, **hyponym** *Bacterium coli commune* Escherich 1885)	15	**8**:73–74 (1958)
Gallionella Ehrenberg 1838	*Gallionella ferruginea* Ehrenberg 1838	9	4:146–147 (1954)
Klebsiella Trevisan 1885	*Klebsiella pneumoniae* (Schroeter) Trevisan 1887 (*Bacterium pneumoniae-crouposae* Zopf 1885)	13	4:151–156 (1954)
Kurthia Trevisan 1885	*Kurthia zopfii* (Kurth) Trevisan 1885 (*Bacterium zopfii* Kurth 1883)	13	4:151–156 (1954)
Lactobacillus Beijerinck 1901	*Lactobacillus delbrueckii* Beijerinck 1901 (not *Lactobacillus caucasicus* Beijerinck 1901)	38	**21**:104 (1971)

LIST 2. *Conserved names of genera of bacteria (nomina generum bacteriorum conservanda)* (continued)

Conserved generic names (*nomina generum conservanda*)	Name of type species of conserved genus	Opinion no.	Citations
Leptotrichia Trevisan 1879	*Leptotrichia buccalis* (Robin) Trevisan 1879 (*Leptothrix buccalis* Robin 1853)	13	**4**:151–156 (1954)
Listeria Pirie 1940	*Listeria monocytogenes* (Murray, Webb and Swann) Pirie 1940 (*Bacterium monocytogenes* Murray, Webb and Swann 1926)	12	**4**:150–151 (1954)
Moraxella Lwoff 1939	*Moraxella lacunata* (Eyre) Lwoff 1939	41	**21**:106 (1971)
Mycoplasma Nowak 1929	*Mycoplasma mycoides* (Borrel *et al.*) Freundt 1955	22	**8**:166–168 (1958)
Neisseria Trevisan 1885	*Neisseria gonorrhoeae* (Zopf) Trevisan 1885 (*Merismopedia gonorrhoeae* Zopf 1885)	13	**4**:151–156 (1954)
Nitrobacter Winogradsky 1892	*Nitrobacter winogradskyi* Winslow *et al.* 1917	23	**8**:169–170 (1958)
Nitrosococcus Winogradsky 1892	*Nitrosococcus nitrosus* (Migula) Buchanan 1925	23	**8**:169–170 (1958)
Nitrosomonas Winogradsky 1892	*Nitrosomonas europaea* Winogradsky 1892	23	**8**:169–170 (1958)
Nocardia Trevisan 1889	*Nocardia farcinica* Trevisan 1889	13	**3**:87–100 (1953) **3**:141–154 (1953) **4**:151–156 (1954)
Pasteurella Trevisan 1887	*Pasteurella cholerae-gallinarum* Trevisan 1887 (*Micrococcus cholerae-gallinarum* Zopf 1885)	13	**4**:151–156 (1954)
Pseudomonas Migula 1894	*Pseudomonas aeruginosa* (Schroeter) Migula 1900 (*Bacterium aeruginosum* Schroeter 1872)	5	**2**:121–122 (1952)
Rhizobium Frank 1889	*Rhizobium leguminosarum* (Frank) Frank 1889 (*Schinzia leguminosarum* Frank1879)	34	**20**:11–12 (1970)

LIST 2. *Conserved names of genera of bacteria (nomina generum bacteriorum conservanda)* (continued)

Conserved generic names (*nomina generum conservanda*)	Name of type species of conserved genus	Opinion no.	Citations
Rickettsia da Rocha-Lima 1916	*Rickettsia prowazekii* da Rocha-Lima 1916	19	**8**:158–159 (1958)
Rhodopseudomonas Czurda and Maresch *emend.* van Niel 1944	*Rhodopseudomonas palustris* (Molisch) van Niel 1944 (*Rhodobacillus palustris* Molisch 1907)	49	**24**:551 (1974)
Selenomonas von Prowazek 1913	*Selenomonas sputigena* (Flügge) Boskamp 1922 (basonym *Spirillum sputigenum* Flügge 1886)	21	**8**:163–165 (1958)
Staphylococcus Rosenbach 1884	*Staphylococcus aureus* Rosenbach 1884	17	**8**:153–154 (1958)
Vibrio Pacini 1854	*Vibrio cholerae* Pacini 1854	31	**15**:185–186 (1965)

LIST 3. *Conserved specific epithets in names of species of bacteria (epitheta specifica conservanda)*

Conserved specific epithets (*epitheta specifica conservanda*)	Name of species in which specific epithet is conserved	Opinion no.	Citations
agalactiae	*Streptococcus agalactiae* Lehmann and Neumann 1896 (*Streptococcus agalactiae contagiosae* Kitt 1893)	8	**4**:145–146 (1954)
avium	*Mycobacterium avium* Chester 1901	47	**23**:472 (1973)
boydii	*Shigella boydii* Ewing 1949	11	**4**:148–150 (1954)
cholerae	*Vibrio cholerae* Pacini 1854	31	**15**:185–186 (1965)
faecalis	*Streptococcus faecalis* Andrewes and Horder 1906	30	**13**:167 (1963)
fermentum	*Lactobacillus fermentum* Beijerinck 1901	50	**24**:551–552 (1974)

List 3. *Conserved specific epithets in names of species of bacteria (epitheta specifica conservanda)* (continued)

Conserved specific epithets *(epitheta specifica conservanda)*	Name of species in which specific epithet is conserved	Opinion no.	Citations
flexneri	*Shigella flexneri* Castellani and Chalmers 1919 (*Bacillus dysenteriae* Flexner 1900)	11	4:148–150 (1954)
fortuitum	*Mycobacterium fortuitum* da Costa Cruz 1938	51	**24**:552 (1974)
meningitidis	The meningococcus (*Diplococcus intracellularis meningitidis* Weichselbaum 1887)	35	**20**:13–14 (1970)
phenylpyruvica	*Moraxella phenylpyruvica* Bøvre and Henriksen 1967	42	**21**:107 (1971)
prowazekii	*Rickettsia prowazekii* da Rocha-Lima 1916	19	**8**:158–159 (1958)
rhusiopathiae	*Erysipelothrix rhusiopathiae* (Migula) Buchanan 1918	32	**20**:9 (1970)
sonnei	*Shigella sonnei* (Levine) Weldin 1927 (*Bacterium sonnei* Levine 1920)	11	4:148–150 (1954)
sphaeroides	*Rhodopseudomonas sphaeroides* van Niel 1944	43	**21**:108 (1971)
typhi	*Salmonella typhi* (Schroeter) Warren and Scott 1930 (*Bacillus typhi* Schroeter 1886)	18	**13**:31–33 (1963) see also **8**:155–156 (1958)

LIST 4. *Rejected names of genera and subgenera of bacteria (nomina generum et subgenerum bacteriorum rejicienda)*

Rejected generic or subgeneric names *(nomina generum et subgenerum rejicienda)*	Names of type species of rejected genera or subgenera	Notes	Opinion no.	Citations
Aerobacter Beijerinck 1900	*Aerobacter aerogenes* (Kruse) Beijerinck 1900	*Nomen ambiguum*	46	**21**:110 (1971)
Astasia Meyer 1897	*Astasia asterospora* Meyer 1897	Later homonym of *Astasia* Ehrenberg 1830 (*Protozoa*)	14	**4**:156–158 (1954)
Astasia Pribram 1929	None named. No species listed.	Later homonym of protozoan generic name *Astasia* Ehrenberg 1830	14	**4**:156–158 (1954)
Babesia Trevisan 1889	*Babesia xanthopyretica* (sic) Trevisan 1880 (*Streptococcus xanthopyreticus* Trevisan 1887)	The later homonym *Babesia* Starcovici 1893 is in common use as the name of a protozoan genus. *Nomen confusum.*	13	**4**:151–156 (1954)
Bacteriopsis Trevisan 1885 (subgenus)	*Bacteriopsis rasmusseni* Trevisan 1885 (*Leptothrix* I Rasmussen 1883)	*Nomen confusum*	13	**4**:151–156 (1954)
Bacterium Ehrenberg 1828	*Bacterium triloculare* Ehrenberg 1828	*Nomen dubium*	4 (revised)	**4**:142 (1954) see also **1**:145–146 (1951) and **3**:141–154 (1953)
Billetia Trevisan 1889	*Billetia laminariae* (Billet) Trevisan 1889 (*Bacterium laminariae* Billet 1888)	*Nomen dubium*	13	**4**:151–156 (1954)

LIST 4. *Rejected names of genera and subgenera of bacteria (nomina generum et subgenerum bacteriorum rejicienda)* (continued)

Rejected generic or subgeneric names *(nomina generum et subgenerum rejicienda)*	Names of type species of rejected genera or subgenera	Notes	Opinion no.	Citations
Castellanella Pacheco and Rodrigues 1930	*Castellanella alcalescens* (Andrewes) Pacheco and Rodrigues 1930 (*Bacillus alkalescens* Andrewes 1918)	Illegitimate later homonym of *Castellanella* Chalmers 1918 (*Protozoa*)	14	4:156–158 (1954)
Cenomesia Trevisan 1889	*Cenomesia albida* Trevisan 1889	*Nomen dubium*	13	4:151–156 (1954)
Chlorobacterium Guillebeau 1890	*Chlorobacterium lactis* Guillebeau 1890		6	4:143 (1954)
Chromobacterium Bergonzini 1879	None designated		16	8:151–152 (1958)
Cloaca Castellani and Chalmers 1919	*Cloaca cloacae* (Jordan) Castellani and Chalmers 1919		28	13:38 (1963)
Coccomonas Orla-Jensen 1921	None. No species included.	Later illegitimate homonym of *Coccomonas* Stein 1878 (*Protozoa*)	14	4:156–158 (1954)
Cornilia Trevisan 1889	*Cornilia alvei* (Cheshire and Cheyne) Trevisan 1889 (*Bacillus alvei* Cheshire and Cheyne 1885)		13	4:151–156 (1954)
Dicoccia Trevisan 1889	*Dicoccia glossophila* Trevisan 1889		13	4:151–156 (1954)
Eucornilia Trevisan 1889 (subgenus)	*Cornilia (Eucornilia) alvei* (Cheshire and Cheyne) Trevisan 1889 (*Bacillus alvei* Cheshire and Cheyne 1885)		13	4:151–156 (1954)

LIST 4. *Rejected names of genera and subgenera of bacteria (nomina generum et subgenerum bacteriorum rejicienda)* (continued)

Rejected generic or subgeneric names *(nomina generum et subgenerum rejicienda)*	Names of type species of rejected genera or subgenera	Notes	Opinion no.	Citations
Eumantegazzaea Trevisan 1889 (subgenus)	*Mantegazzaea (Eumantegazzaea) cienkowskii* Trevisan 1889	*Nomen dubium*	13	4:151–156 (1954)
Eupacinia Trevisan 1889 (subgenus)	*Pacinia (Eupacinia) putrifica* Trevisan 1889 (*Bacillus putrificus coli* Flügge 1886)	*Nomen confusum*	13	4:151–156 (1954)
Euspirillum Trevisan 1889 (subgenus)	*Spirillum (Euspirillum) undula* (Mueller) Ehrenberg 1830 (*Vibrio undula* Mueller 1773)		13	4:151–156 (1954)
Gaffkya Trevisan 1885	*Gaffkya tetragena* (Gaffky) Trevisan 1885		39	**21**:104–105 (1971)
Herellea De Bord 1942	*Herellea vaginicola* De Bord 1942		40	**21**:105–106 (1971)
Leptotrichiella Trevisan 1889 (subgenus)	*Leptotrichia (Leptotrichiella) amphibola* Trevisan 1889	*Nomen dubium*	13	4:151–156 (1954)
Listerella Pirie 1927	*Listerella hepatolytica* Pirie 1927 (*Bacterium monocytogenes* Murray, Webb and Swann 1926)	Illegitimate later homonym of *Listerella* Jahn 1906 (*Myxomycetes*)	14	4:156–158 (1954)
Mantegazzaea Trevisan 1879	*Mantegazzaea cienkowskii* Trevisan 1879	*Nomen dubium*	13	4:151–156 (1954)
Mima De Bord 1939, 1942	*Mima polymorpha* De Bord 1939, 1942		40	**21**:105–106 (1971)

LIST 4. *Rejected names of genera and subgenera of bacteria (nomina generum et subgenerum bacteriorum rejicienda)* (continued)

Rejected generic or subgeneric names *(nomina generum et subgenerum rejicienda)*	Names of type species of rejected genera or subgenera	Notes	Opinion no.	Citations
Nitromonas Winogradsky 1890	None designated		23	8:169–170 (1958)
Nitromonas Orla-Jensen 1909	None designated		23	8:169–170 (1958)
Octopsis Trevisan 1885	*Octopsis cholerae-gallinarum* Trevisan 1885 (*Micrococcus cholerae-gallinarum* Zopf 1885)		13	4:151–156 (1954)
Palmula Prévot 1938	*Palmula spermoides* (Ninni) Prévot 1938	Illegitimate later homonym of *Palmula* Lea 1833 (*Protozoa*)	14	4:156–158 (1954)
Perroncitoa Trevisan 1889	*Perroncitoa scarlatinosa* (Trevisan) Trevisan 1889 (*Micrococcus scarlatinosus* Trevisan 1879)	*Nomen dubium*	13	4:151–156 (1954)
Pfeifferella Buchanan 1918	*Pfeifferella mallei* (Zopf) Buchanan 1918 (*Bacillus mallei* Zopf 1885)	Illegitimate later homonym of *Pfeifferella* Labbé 1899 (*Protozoa*)	14	4:156–158 (1954)
Phytomonas Bergey *et al.* 1923	*Phytomonas campestris* (Pammel) Bergey *et al.* 1923 (*Bacillus campestris* Pammel 1895)	Illegitimate later homonym of *Phytomonas* Donovan 1909 (*Protozoa*)	14	4:156–158 (1954)

LIST 4. *Rejected names of genera and subgenera of bacteria (nomina generum et subgenerum bacteriorum rejicienda)* (continued)

Rejected generic or subgeneric names (*nomina generum et subgenerum rejicienda*)	Names of type species of rejected genera or subgenera	Notes	Opinion no.	Citations
Pleurospora Trevisan 1889 (subgenus)	*Cornilia (Pleurospora) tremula* (Koch) Trevisan 1889 (*Bacillus tremulus* Koch 1877)	*Nomen dubium*	13	4:151–156 (1954)
Polymonas Lieske 1928	*Polymonas tumefaciens* (Smith and Townsend) Lieske 1928 (*Bacterium tumefaciens* Smith and Townsend 1907)		33	**20**:10 (1970)
Pseudospira Trevisan 1889 (subgenus)	*Pacinia (Pseudospira) cholerae-asiaticae* Trevisan 1889		13	4:151–156 (1954)
Pseudospirillum Trevisan 1889 (subgenus)	*Spirillum (Pseudospirillum) amphibolum* Trevisan 1889	*Nomen dubium*	13	4:151–156 (1954)
Rhizomonas Orla-Jensen 1909	None. No species included	Later homonym of *Rhizomonas* Kent 1880 (*Protozoa*)	14	4:156–158 (1954)
Rhodosphaera Buchanan 1918	*Rhodosphaera capsulata* (Molisch) Buchanan 1918 (*Rhodoccocus capsulatus* Molisch 1907)	Later homonym of *Rhodosphaera* Haeckel 1881 (*Protozoa*)	14	4:156–158 (1954)

LIST 5. *Rejected specific epithets in names of species of bacteria (epitheta specifica rejicienda)*

Rejected specific epithets *(epitheta specifica rejicienda)*	Name of species in which specific epithet is rejected	Opinion no.	Citations
caucasicus	*Lactobacillus caucasicus* Beijerinck 1901	38	21:104 (1971)
citrovorum	*Leuconostoc citrovorum* (Hammer) Hucker and Pederson 1931	45	21:109–110 (1971)
liquefaciens	*Aerobacter liquefaciens* Beijerinck 1901	48	23:473–474 (1973)
polymorpha	*Mima polymorpha* De Bord 1939, 1942	40	21:105–106 (1971)
vaginicola	*Herellea vaginicola* De Bord 1942	40	21:105–106 (1971)

Appendix 5

Opinions Relating to the Nomenclature of Bacteria

Appendix 5. Opinions Relating to the Nomenclature of Bacteria

LIST OF OPINIONS

Opinions issued by the International Committee on Bacteriological Nomenclature at the Second International Congress for Microbiology, London, 1936

Opinion	Title	Reference and notes	Result
A	Conservation of the generic name *Bacillus* Cohn 1872, designation of the type species, and of the type strain of the species	Journal of Bacteriology **33**:445–447 (1937), and International Code of Nomenclature of Bacteria and Viruses (1958), p. 148	(a) It was agreed that *Bacillus* Cohn 1872 should be designated as a *genus conservandum*. (b) It was agreed that the type species of *Bacillus* should be designated as *Bacillus subtilis* Cohn 1872 *emendavit* Prazmowski 1880. (c) It was agreed that the type (or standard) strain should be the Marburg strain. (d) It was agreed that cultures of the type (or standard) strain of *Bacillus subtilis* together with complete description should be maintained at each of the recognized Type Culture Collections. (e) It was agreed that the genus *Bacillus* should be so defined as to exclude bacterial species which do not produce endospores. (f) It was agreed that the term *Bacillus* should be used as a generic name and that it should be differentiated from the terms "bacillus," "bacille," and "Bazillus" used as morphological designations.
B	Generic homonyms in the group *Protista*	Journal of Bacteriology **33**:445–447 (1937), and International Code of Nomenclature of Bacteria and Viruses (1958), p. 148	(a) It was agreed that generic homonyms are not permitted in the group *Protista*. (b) It was agreed that it is advisable to avoid homonyms amongst *Protista* on the one hand, and a plant or animal on the other.

LIST OF OPINIONS

Opinions issued by the International Committee on Bacteriological Nomenclature at the Second International Congress for Microbiology, London, 1936 (continued)

Opinion	Title	Reference and notes	Result
C	Capitalization of specific epithets derived from names of persons	Journal of Bacteriology 33:445–447 (1937), and International Code of Nomenclature of Bacteria and Viruses (1958), p. 148	It was agreed that while specific substantive names derived from names of persons may be written with a capital initial letter, all other specific names are to be written with a small initial letter. *Note.* This Opinion is revoked by Rule 59 of this Code, and Recommendation 27h of the 1958 and 1966 editions of the *International Code of Nomenclature of Bacteria (and Viruses)* stated: "A specific epithet, even one derived from the name of a person, should not be written with an initial capital letter."

LIST OF OPINIONS

Opinions issued by the Judicial Commission

Opinion	Title	Reference and notes[a]	Result
1	The correct spelling of the specific epithet in the species name *Bacillus megaterium* de Bary 1884	1 (Part 1) :35–36 (1951)	The spelling *megaterium* of the specific epithet in *Bacillus megaterium* de Bary is to be preferred to the spelling *megatherium.*
2	The combining forms (stems) of compound bacterial generic names ending in *-bacterium, -bacter,* or *-bactrum* (*-bactron*)	1 (Part 1) :37–38 (1951)	The combining form or stem of the last component of names ending in *-bacterium* is *-bacteri,* of those ending in *-bactrum* or *bactron* is *-bactr,* and of those ending in *-bacter* is *-bacter.* Family names derived from such generic names have respectively the endings *-bacteriaceae, -bactraceae* and *-bacteraceae.*
3	Gender of bacterial names ending in *-bacter*	1 (Part 2) :36–37 (1951) ; 1:84–85 in re-issue of volume (1951)	The names of bacterial genera which end in *-bacter* should be regarded as having the masculine gender.

[a] The references are to volumes and pages in the *International Bulletin of Bacteriological Nomenclature and Taxonomy,* to vol. 15 (1965) and thereafter the *International Journal of Systematic Bacteriology,* and date.

LIST OF OPINIONS

Opinions issued by the Judicial Commission

Opinion	Title	Reference and notes	Result
4 (revised)	Rejection of generic name *Bacterium* Ehrenberg	4:142 (1954), see also 1:145–146 (1951) and 3:141–154 Minute 9 (1953)	1. The bacterial generic name *Bacterium* Ehrenberg 1828 is to be recognized as a *nomen generum rejiciendum* (rejected generic name). 2. The bacterial family name *Bacteriaceae* is to be recognized as a *nomen familiae rejiciendum* (rejected family name).
5	Conservation of the generic name *Pseudomonas* Migula 1894 and designation of *Pseudomonas aeruginosa* (Schroeter) Migula 1900 as type species	2:121–122 (1952)	1. The generic name *Pseudomonas* Migula 1894 is to be conserved and placed in the list of *nomina generum conservanda*. 2. The generic name *Pseudomonas* Migula 1894 is to be associated with the species designated and described by Migula 1895. 3. The type species of the genus *Pseudomonas* Migula 1894 is *Pseudomonas aeruginosa* (Schroeter) Migula 1900 (*Bacterium aeruginosum* Schroeter 1872, *Bacillus pyocyaneus* Gessard 1882, *Pseudomonas pyocyanea* Migula 1895).
6	Conservation of the generic name *Chlorobacterium* Lauterborn 1916 against *Chlorobacterium* Guillebeau 1890	4:143 (1954)	The bacterial generic name *Chlorobacterium* Lauterborn 1916 is conserved against the earlier homonym *Chlorobacterium* Guillebeau 1890. The generic name *Chlorobacterium* Guillebeau 1890 is placed in the list of *nomina generum rejicienda*.
7	Nomenclature of the organism associated with granuloma venereum	4:144 (1954), synonymy of *Calymmatobacterium granulomatis* Aragão and Vianna 1913	The bacterial species names *Encapsulatus inguinalis* Bergey *et al.* 1923, *Klebsiella granulomatis* Bergey *et al.* 1925, *Donovania granulomatis* Anderson, de Monbreun and Goodpasture 1944 are later synonyms of *Calymmatobacterium granulomatis* Aragão and Vianna 1913.
8	The correct species name of the streptococcus of bovine mastitis	4:145–146 (1954), conservation of the specific epithet *agalactiae* in the combination *Streptococcus agalactiae* Lehmann and Neumann 1896	The species name *Streptococcus agalactiae* Lehmann and Neumann 1896 is conserved against all synonyms having priority.

LIST OF OPINIONS

Opinions issued by the Judicial Commission

Opinion	Title	Reference and notes	Result
9	Conservation of the bacterial generic name *Gallionella*	4:146–147 (1954), conservation of *Gallionella* Ehrenberg 1838, with type species *Gallionella ferruginea* Ehrenberg	*Gallionella* Ehrenberg is placed in the list of conserved names of bacterial genera (*nomina generum conservanda*) with the type species *Gallionella ferruginea* Ehrenberg.
10	Invalidity of the bacterial generic name *Müllerina* de Petschenko 1910 and of the species name *Müllerina paramecii*	4:147–148 (1954), and status of *Drepanospira* de Petschenko 1911 and *Drepanospira muelleri* de Petschenko 1911	The generic name *Müllerina* de Petschenko 1910 and the species name *Müllerina paramecii* de Petschenko 1910 were not accepted by the author, hence were not validly published and are without standing in nomenclature. The later names *Drepanospira* de Petschenko 1911 and *Drepanospira muelleri* de Petschenko 1911 were validly published and are not later synonyms.
11	Nomenclature of species in the bacterial genus *Shigella*	4:148–150 (1954), validity of publication of the names *Shigella dysenteriae* (Shiga) Castellani and Chalmers 1919, and conservation of the specific epithets *flexneri, boydii* and *sonnei* in, respectively, the species names *Shigella flexneri* Castellani and Chalmers 1919, *Shigella boydii* Ewing 1949 and *Shigella sonnei* (Levine) Weldin 1927, and emendation **10**:85 (1960); **13**:31 (1963)	1. *Shigella dysenteriae* (Shiga) Castellani and Chalmers 1919 was validly published and is legitimate as the name of the dysentery bacterium described by Shiga (1898). 2. The specific epithet *flexneri* in the species name *Shigella flexneri* Castellani and Chalmers 1919 is designated as a conserved specific epithet (*epitheton specificum conservandum*) for the species first described as *Bacillus dysenteriae* Flexner 1900. 3. The species name *Shigella boydii* Ewing 1949 was validly published and is legitimate. The specific epithet *boydii* in the species name *Shigella boydii* is to be conserved (*epitheton specificum conservandum*). 4. The species name *Shigella sonnei* (Levine) Weldin 1927 was validly published and is legitimate. The specific epithet *sonnei* in the species name *Shigella sonnei* is to be conserved (*epitheton specificum conservandum*). 5. A type or standard culture is to be designated by the *Enterobacteriaceae* Subcommittee on Bacteriological Nomenclature for each of the four species. Such cultures as far as possible

			shall be maintained in each of the national Type Culture Collections and in the International Shigella Center, Chamblee, Georgia, U.S.A. (*now in the Center for Disease Control, Atlanta, Georgia*). 6. A culture belonging to the species *Shigella dysenteriae, Shigella flexneri, Shigella boydii,* or *Shigella sonnei* may be completely identified by appending the appropriate serotype number (arabic) to the name.
12	Conservation of *Listeria* Pirie 1940 as a generic name in bacteriology	4:150–151 (1954), type species *Listeria monocytogenes* (Murray, Webb, and Swann) Pirie 1940	*Listeria* Pirie 1940 (type species *Listeria monocytogenes* (Murray, Webb, and Swann) Pirie 1940) shall be placed in the list of conserved names of bacterial genera (*nomina generum conservanda*).
13	Conservation and rejection of names of genera of bacteria proposed by Trevisan 1842–1890	4:151–156 (1954), conservation of generic names *Beggiatoa, Klebsiella, Kurthia, Leptotrichia, Neisseria, Nocardia, Pasteurella*; rejection of generic names *Babesia, Bacteriopsis, Billetia, Cenomesia, Cornilia, Dicoccia, Eucornilia, Eumantegazzaea, Eupacinia, Euspirillum, Leptotrichiella, Mantegazzaea, Octopsis, Perroncitoa, Pleurospora, Pseudospira, Pseudospirillum*; illegitimate generic names *Bollingera, Rasmussenia, Schuetzia, Winogradskya*; of indeterminate status, *Gaffkya, Pacinia*	1. Generic names proposed by Trevisan placed in the list of conserved generic names (*nomina generum conservanda*).

Names of genera and subgenera	Type species
Beggiatoa Trevisan 1842 (p. 56)	*Beggiatoa alba* (Vaucher) Trevisan 1845 (*Oscillaria alba* Vaucher 1803)
Klebsiella Trevisan 1885 (p. 105)	*Klebsiella pneumoniae* (Schroeter) Trevisan 1887 (*Bacterium pneumoniae crouposae* Zopf 1885)
Kurthia Trevisan 1885 (p. 92)	*Kurthia zopfii* (Kurth) Trevisan 1885 (*Bacterium zopfii* Kurth 1883)
Leptotrichia Trevisan 1879 (p. 138)	*Leptotrichia buccalis* (Robin) Trevisan 1879 (*Leptothrix buccalis* Robin 1853)
Neisseria Trevisan 1885 (p. 105)	*Neisseria gonorrhoeae* Trevisan 1885
Nocardia Trevisan 1889 (p. 9)	*Nocardia farcinica* Trevisan 1889

This generic name was omitted in error in the published Opin-

LIST OF OPINIONS

Opinions issued by the Judicial Commission

Opinion	Title	Reference and notes	Result
		ion, and authority is 3:141–154 (1953, Minute 7 File 56) and 3:87–100 (1953).	
		Pasteurella Trevisan 1887 (p. 94)	*Pasteurella cholerae-gallinarum* Trevisan 1887
		2. Generic names proposed by Trevisan placed in the list of rejected generic names (*nomina generum rejicienda*).	
		Babesia Trevisan 1889 (p. 29)	*Babesia xanthopyretica* Trevisan 1889 (*Streptococcus xanthopyreticus* Trevisan 1887)
		Bacteriopsis Trevisan 1885 (p. 103)	*Bacteriopsis rasmussenii* Trevisan 1885 (*Leptothrix* I Rasmussen 1883)
		Billetia Trevisan 1889 (p. 11)	*Billetia laminariae* (Billet) Trevisan 1889 (*Bacterium laminariae* Billet 1888)
		Cenomesia Trevisan 1889 (p. 1039)	*Cenomesia albida* Trevisan 1889
		Cornilia Trevisan 1889 (p. 21)	*Cornilia alvei* (Flügge) Trevisan 1889 (*Bacillus alvei* Flügge 1886)
		Dicoccia Trevisan 1889 (p. 26)	*Dicoccia glossophila* Trevisan 1889
		Eucornilia Trevisan 1889 (p. 21) (Subgenus)	*Cornilia (Eucornilia) alvei* Trevisan 1889 (*Bacillus alvei* Cheshire and Cheyne 1885)
		Eumantegazzaea Trevisan 1889 (p. 942) (Subgenus)	*Mantegazzaea (Eumantegazzaea) cienkowskii* Trevisan 1879
		Eupacinia Trevisan 1889 (p. 23) (Subgenus)	*Pacinia (Eupacinia) putrifica* Trevisan 1889 (*Bacillus putrificus coli* Flügge 1886)

Euspirillum Trevisan 1889 (p. 24) (Subgenus)	*Spirillum (Euspirillum) undula* (Mueller) Ehrenberg 1830 (*Vibrio undula* Mueller 1773)
Leptotrichiella Trevisan 1889 (p. 935) (Subgenus)	*Leptotrichia (Leptotrichiella) amphibola* Trevisan 1889
Mantegazzaea Trevisan 1879 (p. 137)	*Mantegazzaea cienkowskii* Trevisan 1879
Octopsis Trevisan 1885 (p. 102)	*Octopsis cholerae-gallinarum* Trevisan 1885 (*Micrococcus cholerae-gallinarum* Zopf 1885)
Perroncitoa Trevisan 1889 (p. 29)	*Perroncitoa scarlatinosa* (Trevisan) Trevisan 1889 (*Micrococcus scarlatinosus* Trevisan 1879)
Pleurospora Trevisan 1889 (p. 22) (Subgenus)	*Cornilia (Pleurospora) tremula* (Koch) Trevisan 1889 (*Bacillus tremulus* Koch 1877)
Pseudospira Trevisan 1889 (p. 23) (Subgenus)	*Pacinia (Pseudospira) cholerae-asiaticae* Trevisan 1885 (*Vibrio cholerae* Pacini 1854)
Pseudospirillum Trevisan 1889 (p. 25) (Subgenus)	*Spirillum (Pseudospirillum) amphibolum* Trevisan 1889

3. Trevisan's generic names which, as later homonyms or synonyms, are regarded as illegitimate.

Bollingera Trevisan 1889 (p. 26)	*Bollingera equi* (Rivolta) Trevisan (1889) (*Zoogloea pulmonis equi* Bollinger 1870)
Rasmussenia Trevisan 1889 (p. 930)	*Rasmussenia buccalis* (Robin) Trevisan 1889 (*Leptothrix buccalis* Robin 1853)
Schuetzia Trevisan 1889 (p. 29)	*Schuetzia poelsii* Trevisan 1889 (*Streptococcus equi* Sand and Jensen 1888)
Winogradskya Trevisan 1889 (p. 12)	*Winogradskya ramigera* (Itzigsohn) (Trevisan 1889 (*Zoogloea ramigera* Itzigsohn 1867)

LIST OF OPINIONS

Opinions issued by the Judicial Commission

Opinion	Title	Reference and notes	Result
			4. Trevisan's generic names whose status is indeterminate.
			Gaffkya Trevisan 1885 (p. 105); but see Opinion 39 — *Gaffkya tetragena* (Gaffky) Trevisan 1885 (*Micrococcus tetragenus* Gaffky 1883) *Pacinia* Trevisan 1885 (p. 83) ; but see Opinion 31 — *Pacinia cholerae-asiaticae* Trevisan 1885
14	Names of bacterial genera to be rejected as later synonyms of names of genera of protozoa	4:156–158 (1954), rejection of *Astasia* Meyer 1897, *Astasia* Pribram 1929, *Castellanella* Pacheco and Rodrigues 1930, *Charon* Holmes 1948, *Coccomonas* Orla-Jensen 1921, *Listerella* Pirie, 1927, *Palmula* Prévot 1938, *Pfeifferella* Buchanan 1918, *Phytomonas* Bergey *et al.* 1923, *Rhizomonas* Orla-Jensen 1909, *Rhodosphaera* Buchanan 1918	The following names proposed for bacterial genera are found to be later homonyms of names applied to genera of protozoa. Rule 24 of the International Code of Nomenclature of Bacteria and Viruses states that such later homonyms are illegitimate in bacteriology. These names are to be placed in the list of names of bacterial genera to be rejected (*nomina generum bacteriorum rejicienda*) .

Rejected names of bacterial genera	Names of protozoan genera having priority
Astasia Meyer 1897 *Astasia* Pribram 1929	*Astasia* Ehrenberg 1830
Castellanella Pacheco and	*Castellanella* Chalmers 1918

Rodrigues 1930	
Charon Holmes 1948 (a genus of viruses)	*Charon* Karsch 1879
Coccomonas Orla-Jensen 1921	*Coccomonas* Stein 1878
Listerella Pirie 1927	*Listerella* Jahn 1906
Palmula Prévot 1938	*Palmula* Lea 1833
Pfeifferella Buchanan 1918	*Pfeifferella* Labbé 1899
Phytomonas Bergey *et al.* 1923	*Phytomonas* Donovan 1909
Rhizomonas Orla-Jensen 1909	*Rhizomonas* Kent 1880
Rhodosphaera Buchanan 1918	*Rhodosphaera* Haeckel 1881

15	Conservation of the family name *Enterobacteriaceae*, of the name of the type genus, and designation of the type species	**8**:73–74 (1958), with type genus *Escherichia* Castellani and Chalmers 1919 as conserved generic name and type species *Escherichia coli* (Migula) Castellani and Chalmers 1919	1. The family name *Enterobacteriaceae* Rahn 1937 (p. 280) is placed in the list of conserved family names *(nomina conservanda familiarum)*. 2. The genus *Escherichia* Castellani and Chalmers 1919 (p. 941) is designated as the type genus of the family *Enterobacteriaceae* Rahn 1937. 3. The generic name *Escherichia* Castellani and Chalmers 1919 (p. 941) is placed in the list of conserved generic names (*nomina generum conservanda*). 4. The type species of the genus *Escherichia* Castellani and Chalmers 1919 (p. 941) is *Escherichia coli* (Migula) Castellani and Chalmers 1919 (p. 941), basonym: *Bacillus coli* (Migula 1895 (p. 27); hyponym: *Bacterium coli commune* Escherich 1885 (p. 518)).
16	Conservation of the generic name *Chromobacterium* Bergonzini 1880 and designation of the type species and the neotype culture of the type species	**8**:151–152 (1958), type species *Chromobacterium violaceum* Bergonzini 1880	1. The generic name *Chromobacterium* Bergonzini 1879 is rejected and placed in the list of *nomina generum rejicienda.* 2. The generic name *Chromobacterium* Bergonzini 1880 is conserved and placed in the list of *nomina generum conservanda.* 3. The type species of the genus *Chromobacterium* Bergonzini 1880 is *Chromobacterium violaceum* Bergonzini 1880.

LIST OF OPINIONS

Opinions issued by the Judicial Commission

Opinion	Title	Reference and notes	Result
			4. A neotype strain of *Chromobacterium violaceum* Bergonzini 1880 is designated and has been deposited in the American Type Culture Collection, Washington D.C. (ATCC 12472) and in the National Collection of Type Cultures, London (NCTC 9757).
17	Conservation of the generic name *Staphylococcus* Rosenbach, designation of *Staphylococcus aureus* as the nomenclature type of the genus *Staphylococcus* Rosenbach, and designation of a neotype culture of *Staphylococcus aureus* Rosenbach	8:153–154 (1958)	1. The generic name *Staphylococcus* Rosenbach 1884 is conserved and placed in the list of *nomina generum conservanda*. 2. *Staphylococcus aureus* Rosenbach 1884 is recognized as the nomenclatural type species of the genus *Staphylococcus* Rosenbach 1884. 3. The strain labelled NCTC 8532 in the National Collection of Type Cultures, London, is designated as the neotype strain of the species *Staphylococcus aureus* Rosenbach 1884.
18	Conservation of *typhi* in the binary combination *Salmonella typhi*	13:31–33 (1963), see also 8:155–156 (1958)	The specific epithet *typhi* in the name of the species *Salmonella typhi* (Schroeter) Warren and Scott is conserved over the specific epithet *typhosa* in the name of the species *Salmonella typhosa* (Zopf) White 1930, with the recognition of *Bacillus typhi* Schroeter 1886 as the basonym.
19	Conservation of the generic name *Rickettsia* da Rocha-Lima and of the species name *Rickettsia prowazekii* da Rocha-Lima	8:158–159 (1958)	The generic name *Rickettsia* da Rocha-Lima is conserved against *Stricheria* Stempell, and the specific epithet *prowazekii* in the species name *Rickettsia prowazekii* da Rocha-Lima is conserved against the specific epithet *jurgensi* first used in the species name *Stricheria jurgensi* Stempell.
20	Status of new generic names of bacteria published without names of included species	8:160–162 (1958)	1. *Name of a hypothetical genus.* A hypothetical genus is one in which no species is described, named, or cited; the existence of the genus is predicated upon the future discovery and description of species as yet unknown. A name applied to a hypothetical genus is not validly published and is to be placed in the list of *nomina rejicienda*.

2. *Name of a "temporary" genus.* A generic name proposed for a genus whose sole function is stated to be to serve as the temporary generic haven for insufficiently described species, which species may be allocated later to an appropriate genus or genera, is to be regarded as not validly published. Such a name may be placed in the list of *nomina rejicienda.*
3. *Name of a new genus with a described species which is neither named nor identified with a previously named species.* A new generic name published in a combined description of a genus and species, without the species being named, without citation of a previously and effectively published description of the species, and without subsequent acceptance of the generic name and naming of the species by a later author, should be regarded as not validly published. Such a generic name may be placed in the list of *nomina rejicienda.*

   However, if a later author has recognized the generic name and has used it with a specific epithet in naming the species described by the first author, particularly if there has been later general acceptance of the name, there may be validation of the generic name as proposed by its author, with the name of the species ascribed to the later author who gave it. Proposals for such validations of names should be made to the Judicial Commission for appropriate action.
4. *Name of a new genus proposed to include one or more previously described and named species, but without simultaneous publication of the new binary combination of generic name and specific epithet.* A published generic name applied to a new genus in which the generic name is not used in a binary combination in naming any species, but in which there is citation of a previously and effectively published description of a species under another name, is to be regarded as validly published and the consequent *combinationes novae* ascribed likewise to the author of the generic name.

LIST OF OPINIONS

Opinions issued by the Judicial Commission

Opinion	Title	Reference and notes	Result
21	Conservation of the generic name *Selenomonas* von Prowazek	**8**:163–165 (1958), with type species *Selenomonas sputigena* (Flügge) Boskamp 1922	1. The generic name *Selenomonas* von Prowazek 1913 was validly published with an accompanying description of the genus. 2. The species *Spirillum sputigenum* Flügge 1886 was characterized and adequate references to description given. The species was assigned to the genus *Selenomonas*. 3. *Selenomonas sputigena* (Flügge) Boskamp 1922 (basonym: *Spirillum sputigenum* Flügge) is designated as the type species of *Selenomonas* von Prowazek. 4. The generic name *Selenomonas* von Prowazek 1913 is placed in the list of *nomina generum conservanda*.
22	Status of the generic name *Asterococcus* and conservation of the generic name *Mycoplasma*	**8**:166–168 (1958), illegitimacy of *Asterococcus* Borrel *et al.* 1910, conservation of *Mycoplasma* Nowak 1929 with type species *Mycoplasma mycoides* (Borrel *et al.*) Freundt 1955	1. The generic name *Asterococcus* Borrel, Dujardin-Beaumetz, Jeantet, and Jouan 1910 is a later homonym of *Asterococcus* Scherffel 1908, and hence illegitimate. 2. The generic name *Mycoplasma* Nowak 1929 is placed in the list of bacterial *nomina generum conservanda* as the first legitimate generic name proposed to replace *Asterococcus* Borrel *et al.* The type species is *Mycoplasma mycoides* (Borrel *et al.*) Freundt 1955 basonym: *Asterococcus mycoides* Borrell *et al.*).
23	Rejection of the generic names *Nitromonas* Winogradsky 1890 and *Nitromonas* Orla-Jensen 1909, conservation of the generic names *Nitrosomonas* Winogradsky 1892, *Nitrosococcus* Winogradsky 1892, and *Nitrobacter* Winogradsky 1892,	**8**:169–170 (1958), type species are respectively *Nitrosomonas europaea* Winogradsky 1892, *Nitrosococcus nitrosus* (Migula) Buchanan 1925, and *Nitrobacter winogradskyi* Winslow *et al.* 1917	1. The generic name *Nitromonas* Winogradsky 1890 is placed in the list of *nomina generum rejicienda*. 2. The generic name *Nitromonas* Orla-Jensen 1909 is a later homonym of *Nitromonas* Winogradsky 1890 and a later synonym of *Nitrobacter* Winogradsky (1892). It is placed in the list of *nomina generum rejicienda*. 3. The generic name *Nitrosomonas* Winogradsky 1892 is placed in the list of *nomina generum conservanda* with *Nitrosomonas europaea* Winogradsky 1892 as the nomenclatural type species.

	and the designation of the type species of these genera		4. The generic name *Nitrosococcus* Winogradsky 1892 is placed in the list of *nomina generum conservanda*, with the species described by Winogradsky and later named *Nitrosococcus nitrosus* (Migula) Buchanan 1925 as the nomenclatural type species. 5. The generic name *Nitrobacter* Winogradsky 1892 is placed in the list of *nomina generum conservanda*, with the species described by Winogradsky and later named *Nitrobacter winogradskyi* Winslow *et al.* 1917 as the nomenclatural type species.
24	Rejection of the generic name *Arthrobacter* Fischer 1895 and conservation of the generic name *Arthrobacter* Conn and Dimmick 1947	**8**:171–172 (1958), conservation was effected though its mention was omitted in the Opinion itself. The title of the Opinion explicitly states that *Arthrobacter* Conn and Dimmick is conserved.	1. The name *Arthrobacter* proposed by Fischer in 1895 as the name of a hypothetical genus of bacteria was not validly published and has no standing in nomenclature. 2. The generic name *Arthrobacter* Conn and Dimmick 1947 was validly published as a *nomen novum*. It is not an emendation of *Arthrobacter* Fischer 1895 nor a later homonym.
25	Rejection of names of bacteria in certain publications of Trécul, Hallier, Billroth, and Ogston	**13**:33–35 (1963)	1. The specific, subgeneric, generic or other names proposed in the several publications listed below were not validly published as names of taxa of bacteria and have no standing in bacteriological nomenclature. These publications are included in the list of Rejected Publications as authorized in Paragraph 8 under "Functions of the Judicial Commission," in Section IV of the International Code of Nomenclature of the Bacteria and Viruses: a) Trécul, A. 1865. Production de plantules amylifères dans les cellules végétales pendant la putréfaction. Chlorophylle cristallisée. C. R. Acad. Sci. Paris **61**:432–436. b_1) Hallier, Ernst. 1866. Die pflanzlichen Parasiten des menschlichen Körpers für Aerzte, Botaniker und Studierende zugleich als Einleitung in das Stadium der niederen Organismen. Leipzig. b_2) Hallier, Ernst. 1868. Mikroskopische Untersuchungen. Zwei neue Untersuchungen über den *Micrococcus*. Flora N.S. **26**:654–657.

LIST OF OPINIONS

Opinions issued by the Judicial Commission

Opinion	Title	Reference and notes	Result
			b_3) Hallier, E. 1868. Mykologische Untersuchungen. III. Untersuchungen der Parasiten beim Tripper, beim weichen Schanker, bei der Syphilis und bei der Rotzkranheit der Pferde. Flora N.S. **26**:289–301. b_4) Hallier, Ernst. 1870. Die Parasiten der Infektionskrankheiten. Z. Parasitenkd. **2**:113–132. c) Billroth, C.A.T. 1874. Untersuchungen über die Vegetationsformen von *Coccobacteria septica*. Berlin. d_1) Ogston, Alex. 1882. Micrococcus poisoning. J. Anat. Physiol. **16**:526–567. d_2) Ogston, Alex. 1883. Micrococcus poisoning (cont.). J. Anat. Physiol. **17**:24–58. 2. Names proposed in the above-listed publications of Trécul, Hallier, Billroth, and Ogston have in some cases been adopted by later authors as the names of bacterial taxa and one or other of the four authors named cited as author. In such cases the name of the taxon is to be ascribed to the first subsequent author whose publication meets the requirements of valid publication as prescribed in the International Code of Nomenclature of Bacteria and Viruses (Rule 11).
26	Designation of neotype strains (cultures) of type species of the bacterial genera *Salmonella*, *Shigella*, *Arizona*, *Escherichia*, *Citrobacter* and *Proteus* of the family *Enterobacteriaceae*	**13**:35–36 (1963), and **14**:57 (1964)	Neotype cultures of *Salmonella cholerae-suis*, *S. typhi-murium*, *Shigella dysenteriae*, *Arizona arizonae*, *Escherichia coli*, *Citrobacter freundii*, and *Proteus vulgaris* were approved.

Name of species	Catalogue no. NCTC London	Catalogue no. ATCC Washington
Salmonella cholerae-suis (Smith) Weldin 1927. Type species of genus *Salmonella* Lignières 1900.	5735	13312

			Salmonella typhi-murium (Loeffler) Castellani and Chalmers 1919.	74	13311
			Shigella dysenteriae (Shiga) Castellani and Chalmers 1919. Type species of genus *Shigella* Castellani and Chalmers 1919.	4837	13313
			Arizona arizonae Kauffmann and Edwards 1952. Type species of genus *Arizona* Kauffmann and Edwards 1952.	8297	13314
			Escherichia coli (Migula) Castellani and Chalmers 1919. Type species of genus *Escherichia* Castellani and Chalmers 1919.	9001	11775
			Citrobacter freundii (Braak) Werkman and Gillen 1932. Type species of genus *Citrobacter* Werkman and Gillen 1932.	9750	8090
			Proteus vulgaris Hauser 1885. Type species of genus *Proteus* Hauser 1885.	4175	13315
27	Designation of the neotype strain of *Streptococcus agalactiae* Lehmann and Neumann	13:37 (1963)	The strain Stableforth G19 is designated as the neotype strain of *Streptococcus agalactiae* Lehmann and Neumann. This neotype strain is catalogued in the National Collection of Type Cultures as NCTC 8181 and in the American Type Culture Collection as ATCC 13813.		
28	Rejection of the bacterial generic name *Cloaca* Castellani and Chalmers and acceptance of *Enterobacter* Hormaeche and Edwards as a bacterial generic name with type species *Enterobacter cloacae* (Jordan) Hormaeche and Edwards	13:38 (1963), conservation was effected by statement in the Summary though omitted in the title and in the Opinion itself.	The generic name *Cloaca* Castellani and Chalmers is rejected and replaced by the generic name *Enterobacter* Hormaeche and Edwards with the type species *Enterobacter cloacae* (Jordan) Hormaeche and Edwards: the basonym is *Bacillus cloacae* Jordan.		

LIST OF OPINIONS

Opinions issued by the Judicial Commission

Opinion	Title	Reference and notes	Result
29	Designation of strain ATCC 3004 (IMRU 3004) as the neotype strain of *Streptomyces albus* (Rossi Doria) Waksman and Henrici	**13**:123–124 (1963)	The strain labelled ATCC 3004 in the American Type Culture Collection, Washington, D.C., and also known as IMRU 3004 (Institute of Microbiology, Rutgers University) is designated as the neotype strain of *Streptomyces albus* (Rossi Doria) Waksman and Henrici 1943.
30	Conservation of the specific epithet *faecalis* in the species name *Streptococcus faecalis* Andrewes and Horder 1906	**13**:167 (1963)	The specific epithet *faecalis* in the species name *Streptococcus faecalis* Andrewes and Horder 1906 is conserved against the specific epithets in *Streptococcus liquefaciens* Sternberg 1892, *S. zymogenes* McCallum and Hastings 1899, and all other earlier synonymous specific epithets in the genus *Streptococcus*.
31	Conservation of *Vibrio* Pacini 1854 as a bacterial generic name, conservation of *Vibrio cholerae* Pacini 1854 as the nomenclatural type species of the bacterial genus *Vibrio*, and designation of neotype strain of *Vibrio cholerae* Pacini	**15**:185–186 (1965)	*Vibrio cholerae* Pacini 1854 is conserved as the name of the type species of the bacterial genus *Vibrio* Pacini 1854, the bacterial generic name *Vibrio* Pacini 1854 is placed in the list of conserved bacterial generic names (*nomina generum conservanda*), and National Collection of Type Cultures NCTC 8021 (American Type Culture Collection, ATCC 14035) is designated as the neotype of the species *Vibrio cholerae* Pacini 1854.
32	Conservation of the specific epithet *rhusiopathiae* in the scientific name of the organism known as *Erysipelothrix rhusiopathiae* (Migula 1900) Buchanan 1918	**20**:9 (1970)	The specific epithet *rhusiopathiae* in the scientific name of the organism known as *Erysipelothrix rhusiopathiae* (Migula 1900) Buchanan 1918 is conserved against the specific epithet *insidiosa* (basonym: *Bacillus insidiosus* Trevisan 1885) and against all other specific epithets applied to this organism.
33	Conservation of the generic name *Agrobacterium* Conn 1942	**20**:10 (1970), type species *Agrobacterium tumefaciens* (Smith and Townsend) Conn 1942	The generic name *Agrobacterium* Conn 1942 is conserved against the name *Polymonas* Lieske 1928, which is placed in the list of *nomina generum rejicienda*. The type species, by original designation, is *Agrobacterium tumefaciens* (Smith and Town-

			send 1907) Conn. 1942; the basonym is *Bacterium tumefaciens* Smith and Townsend 1907.
34	Conservation of the generic name *Rhizobium* Frank 1889	**20**:11–12 (1970), type species *Rhizobium leguminosarum* Frank 1889	The generic name *Rhizobium* Frank 1889 is conserved against *Phytomyxa* Schroeter 1886 and all earlier synonyms. The type species is *Rhizobium leguminosarum* (Frank 1879) Frank 1889; the basonym is *Schinzia leguminosarum* Frank 1879.
35	Conservation of the specific epithet *meningitidis* in the scientific name of the meningococcus	**20**:13–14 (1970), and designation of neotype strain	The specific epithet "*meningitidis*" is conserved in the scientific name of the meningococcus (*Diplococcus intracellularis meningitidis* Weichselbaum) against all earlier specific epithets. The neotype strain of this organism is ATCC 13077 (= Sara E. Branham M1027 = NCTC 10025).
36	Designation of strain ATCC 10145 as the neotype strain of *Pseudomonas aeruginosa* (Schroeter) Migula	**20**:15–16 (1970)	The neotype strain of *Pseudomonas aeruginosa* (Schroeter) Migula is ATCC 10145 = CCEB 481 = IBCS 277 = NCIB 8295 = NCTC 10332 = NRRL B-771 = RH 815.
37	Designation of strain ATCC 13525 as the neotype strain of *Pseudomonas fluorescens* Migula	**20**:17–18 (1970)	The neotype strain of *Pseudomonas fluorescens* Migula is ATCC 13525 = CCEB 546 = NCIB 9046 = NCTC 10038 = RH 818 = M. Rhodes 28/5.
38	Conservation of the generic name *Lactobacillus* Beijerinck	**21**:104 (1971), with new type species *Lactobacillus delbrueckii* Beijerinck 1901 and neotype strain	The generic name *Lactobacillus* Beijerinck 1901 is conserved over *Saccharobacillus* van Laer 1892 and all earlier objective synonyms. The type species of this genus is *Lactobacillus delbrueckii* Beijerinck 1901, the neotype strain of which is ATCC 9649 = NCDO 213. The name *Lactobacillus delbrueckii* Beijerinck 1901, although used by Beijerinck as a simplified version of the subspecific name "*Lactobacillus fermentum* var. *delbrucki*," shall be held to be validly published by Beijerinck as a species name. The name *Lactobacillus caucasicus* Beijerinck 1901 is placed in the list of rejected names, and *L. caucasicus* ceases to be the type species of *Lactobacillus* Beijerinck.
39	Rejection of the generic name *Gaffkya* Trevisan	**21**:104–105 (1971)	The generic name *Gaffkya* Trevisan 1885 is placed on the list of rejected names.

LIST OF OPINIONS

Opinions issued by the Judicial Commission

Opinion	Title	Reference and notes	Result
40	Rejection of the names *Mima* De Bord and *Herellea* De Bord and of the specific epithets *polymorpha* and *vaginicola* in *Mima polymorpha* De Bord and *Herellea vaginicola* De Bord, respectively	**21**:105–106 (1971), and loss of standing in nomenclature of the tribal name *Mimeae* De Bord 1939	The generic names *Mima* De Bord 1939, 1942 and *Herellea* De Bord 1942 are placed on the list of rejected names. The specific epithets *polymorpha* and *vaginicola* in *Mima polymorpha* De Bord 1939, 1942 and *Herellea vaginicola* De Bord 1942 respectively are placed on the list of rejected epithets. The tribal name *Mimeae* De Bord 1939, 1942 therefore loses its standing in nomenclature.
41	Conservation of the generic name *Moraxella* Lwoff	**21**:106 (1971), type species *Moraxella lacunata* (Eyre) Lwoff 1939, and neotype strain	The generic name *Moraxella* Lwoff 1939 is conserved over *Diplobacillus* McNab 1904 and over all earlier objective synonyms. The type species is *Moraxella lacunata* (Eyre) Lwoff 1939, and the neotype strain of this species is Morax 260 = ATCC 17967.
42	Conservation of the specific epithet *"phenylpyruvica"* in the name *Moraxella phenylpyruvica* Bøvre and Henriksen	**21**:107 (1971), conservation over epithet *polymorpha* in the name *Moraxella polymorpha* Flamm 1957, and neotype strain	The specific epithet *"phenylpyruvica"* in the name *Moraxella phenylpyruvica* Bøvre and Henriksen 1967 is conserved against the specific epithet *"polymorpha"* in the name of the earlier objective synonym *Moraxella polymorpha* Flamm 1957 and against the specific epithets in all other earlier objective synonyms. The neotype strain of *Moraxella phenylpyruvica* is 2863 (= ATCC 23333 = NCTC 10526).
43	Conservation of the specific epithet *"sphaeroides"* in the name *Rhodopseudomonas sphaeroides* van Niel	**21**:108 (1971), and neotype strain	The specific epithet *"sphaeroides"* in the name *Rhodopseudomonas sphaeroides* van Niel 1944 is conserved against the specific epithet *"minor"* in the name of the earlier subjective synonym *Rhodococcus minor* and against the specific epithets in the names of all earlier objective synonyms of *Rhodopseudomonas sphaeroides*. The neotype strain is van Niel's ATH 2.4.1 (= ATCC 17023).
44	Validation of the generic name *Chloropseudomonas* Czurda and Maresch 1937 and designation of the type species	**21**:109 (1971), type species *Chloropseudomonas ethylica* Shaposhnikov *et al.* 1960	The generic name *Chloropseudomonas* is held to be validly published by Czurda and Maresch 1937. The type species is *Chloropseudomonas ethylica* Shaposhnikov, Kondratieva and Fedorov 1960.

45	Rejection of the name *Leuconostoc citrovorum* (Hammer) Hucker and Pederson	**21**:109–110 (1971)	The name *Leuconostoc citrovorum* (Hammer 1920) Hucker and Pederson 1931, together with its objective synonyms, is regarded as a *nomen dubium* and is placed on the list of rejected names.
46	Rejection of the generic name *Aerobacter* Beijerinck	**21**:110 (1971)	The generic name *Aerobacter* Beijerinck 1900 is regarded as a *nomen ambiguum* and is placed on the list of rejected generic names.
47	Conservation of the specific epithet *avium* in the scientific name of the agent of avian tuberculosis	**23**:472 (1973)	The specific epithet *avium* is conserved against the specific epithet *tuberculosis-gallinarum* and all earlier objective synonyms in the scientific name of the agent of avian tuberculosis. The name *Mycobacterium avium* shall be held to be validly published by Chester in 1901. The neotype strain of *M. avium* Chester is ATCC 25291.
48	Rejection of the name *Aerobacter liquefaciens* Beijerinck and conservation of the name *Aeromonas* Stanier with *Aeromonas hydrophila* as the type species	**23**:473–474 (1973)	The name *Aerobacter liquefaciens* Beijerinck 1900 is a *nomen dubium* and, together with all objective synonyms of this name, is placed on the list of rejected names. The generic name *Aeromonas* Stanier 1943, with type species *Aeromonas hydrophila* (Chester 1901) Stanier 1943, is conserved. The name *Aeromonas* is not to be attributed to Kluyver and van Niel. The neotype strain of *A. hydrophila* is ATCC 7966.
49	Conservation of the generic name *Rhodopseudomonas* Czurda and Maresch emend. van Niel	**24**:551 (1974)	The generic name *Rhodopseudomonas* Czurda and Maresch 1937 emend. van Niel 1944 is conserved over all earlier objective synonyms; the type species is *Rhodopseudomonas palustris* (Molisch 1907) van Niel 1944 (basonym: *Rhodobacillus palustris* Molisch 1907).
50	Conservation of the epithet *fermentum* in the combination *Lactobacillus fermentum* Beijerinck	**24**:551–552 (1974)	The species name *Lactobacillus fermentum* Beijerinck 1901 shall be held to be validly published by Beijerinck 1901 as the name of a bacterial species, and the epithet *fermentum* in the combination *Lactobacillus fermentum* Beijerinck 1901 is conserved over the epithets in all other objective synonyms. The neotype strain of *Lactobacillus fermentum* is ATCC 14931.
51	Conservation of the epithet *fortuitum* in the combination *Mycobacterium fortuitum* da Costa Cruz	**24**:552 (1974)	The specific epithet *fortuitum* in the name *Mycobacterium fortuitum* da Costa Cruz 1938 is conserved against the epithet *ranae* in the subjective synonym *Mycobacterium ranae* Bergey *et al.* 1923 and against the specific epithets in the names of all objective synonyms of *Mycobacterium fortuitum* and *Mycobacterium ranae*. The type strain of *Mycobacterium fortuitum* is ATCC 6841.

Appendix 6

Published Sources for Recommended Minimal Descriptions

Appendix 6. Published Sources for Recommended Minimal Descriptions

(References to the publications where the recommendations for minimal standards can be found will be included in this appendix in future editions of this Code.)

Appendix 7

Publication of a New Name

Appendix 7. Publication of a New Name

After the date of publication of this revision of the Code, valid publication of the name of a taxon (including a new combination) requires publication in the *International Journal of Systematic Bacteriology* of (a) the name of the taxon, (b) for new taxa the designation of a type, and (c) a description or a reference to an effectively published description of the taxon whether in the *International Journal of Systematic Bacteriology* or in another publication. Fuller details are given below.

(1) The name should be in the correct form. Generic and suprageneric names are single words in Latin form and spelled with an initial capital letter. Names of species are binary combinations of words in Latin form consisting of a generic name and a single, specific epithet, the latter spelled with an initial lower-case letter. Subspecific names are ternary combinations consisting of the name of a species followed by the term "subspecies" (ordinarily "subsp.") and this in turn by a single subspecific epithet. Names of taxa from the rank of order to tribe inclusive are formed by the addition of the appropriate suffix to the stem of the name of the type genus (see 5 below). The suffix for order is *-ales,* for suborder *-ineae,* for family *-aceae,* for subfamily *-oideae,* for tribe *-eae,* and for subtribe *-inae.*

Although not a requirement for the valid publication of a new name, the derivation of the name should be given.

Where possible, the title of the paper should include any new names or combinations that are proposed in the text.

(2) The name should be clearly proposed as a new name or combination and should be accepted by the author at the time of publication. New names are ordinarily proposed by an author appending the phrase *"species nova"* (abbreviation: sp. nov.), *"genus novum"* (abbreviation: gen. nov.), *"combinatio nova"* (abbreviation: comb. nov.), or the like after the name or combination he is proposing as new; alternatively, the author may make a statement to the effect that he is introducing a new name or combination.

(3) The name should not be a later homonym of a previously validly published name of an alga, bacterium, fungus, protozoon, or virus. (See the *International Journal of Systematic Bacteriology* from 1975 onwards, and possibly Appendix 2 in future editions of this Code, for names of bacterial taxa; and see Appendix 3 for published sources of names of algal, protozoal, fungal, and viral taxa.)

(4) The name must be accompanied by a description of the taxon or

by a reference to a previously published description of the taxon (see [6] below).

(5) The nomenclatural type of a new taxon should be designated. In the case of species and subspecies which can be cultivated, the type strain should be described by itself and should be designated by the author's strain number as well as the accession number under which it is held by at least one culture collection from which cultures of the strain are available.

A nomenclatural type is that constituent element of a taxon to which the name of the taxon is permanently attached. The type of a species or a subspecies is a strain, that of a genus is a species, and that of an order, family, subfamily, tribe, or subtribe is the genus on whose name the name of the higher taxon is based (see [1] above). For species and subspecies whose cells cannot be maintained in culture or for which cultures are not maintained, the type strain can be represented by the original description and by illustrations and specimens.

A type strain is one of the strains on which the author who first described a named organism based his description of the organism and which he or a subsequent author definitely designated as a type; if the description was based on a single strain, this strain is the type by monotypy.

A neotype strain replaces a type strain which can no longer be found. The neotype should possess the characteristics as given in the original description; any deviations should be explained. A neotype strain must be proposed by an author in the *International Journal of Systematic Bacteriology* (*proposed neotype*) together with a reference (or references) to the first description and name for the microorganism (or to an approved list if appropriate), a description (or reference to a description) of the proposed neotype strain, and a record of the author's designation for the type strain and of at least one culture collection from which cultures of the strain are available. The neotype strain becomes established two years after the date of publication in the *International Journal of Systematic Bacteriology* (*established neotype*). Any objections should be referred to the Judicial Commission within the first year after publication of the proposal. A neotype strain shall be proposed only after a careful search for original strains. If an original strain is subsequently discovered, the matter shall be referred immediately to the Judicial Commission. Allowance is made for replacement of a type strain.

(6) Descriptions of taxa should include the following information: (a) those characteristics which are essential for membership in the taxon, i.e., those characteristics which constitute the basic concept of the taxon; (b) those characteristics which qualify the taxon for membership in the next higher taxon; (c) the diagnostic characteristics, i.e., those character-

istics which distinguish the taxon from closely related taxa; and, (d) in the case of species, the total number of strains studied, the strain designations, and the number of strains which are either positive or negative for each characteristic. If the strains are not homogeneous in a characteristic, the specific strain numbers for those strains which disagree with the majority should be given. From this information, the detailed results for each strain can be reconstructed without the full publication of the details for each strain. Where appropriate, suitable photomicrographs and, if necessary, electron photomicrographs should be included as part of the description to show morphological or anatomical characters that are pertinent to the classification. Descriptions should conform at least to such minimal descriptions as have been approved (see Appendix 6).

Appendix 8

Preparation of a Request for an Opinion

Appendix 8. Preparation of a Request for an Opinion

In those cases where strict adherence to the rules of nomenclature would produce confusion or would not result in nomenclatural stability, exceptions to the rules may be requested of the Judicial Commission of the ICSB. Requests for Opinions must be accompanied by a fully documented statement of the relevant facts. The Judicial Commission will consider all Requests for Opinions and either issue an Opinion in the IJSB or signify its attitude in some other way. The title of a manuscript should provide a concise statement of the contents of the manuscript. If an Opinion of the Judicial Commission is requested in the text, "Request for an Opinion" should appear as a subtitle. When a request is not supported by adequate evidence, it will be returned to the author for revision. A Request for an Opinion submitted in an acceptable form will be published as soon as possible in the *International Journal of Systematic Bacteriology*, and microbiologists are invited to submit statements in support of or in opposition to the Request. When an Opinion is challenged, the basis of the challenge should be stated and supported by a documented statement of the relevant facts.

Requests for Opinions or challenges of such Requests or proposals for Opinions or of an issued Opinion should be submitted to the Editorial Secretary in a form suitable for publication without delay in the IJSB.

Appendix 9

Orthography

Appendix 9. Orthography

A. Orthographic Variants

The following are some of the possible orthographic variants found in names and epithets (Table 3).

(1) Different transliterations into Latin of words from other languages.

Examples: *Corynebacterium* and *Corynobacterium.*

(2) Changes in the gender and endings of specific or subspecific epithets to agree with the gender of the generic substantive.

Example: *Micrococcus luteus, Sarcina lutea.*

(3) The use of a wrong or alternative connecting vowel or its omission.

TABLE 3. *Connecting vowels*

Type of compound	Preferred connecting vowel	Orthographic variant	Examples
Latin-Latin	-i-	-o-	*rubrilineans, rubrofuscus*
Greek-Greek	-o- None if first stem ends in -y- Another vowel if precedent is found in Greek	Presence or absence of -o-, or have another combining vowel	*achromogenes* *glycyphyllus,*[a] *ichthyodermis* *halmephilus,* (from ἅλμη, brine)
Two or more languages	No special vowel	A different connecting vowel	*disciformis* (Greek-Latin)

[a] Not in *Index Bergeyana,* so probably hypothetical.

(4) Words of Greek origin differing merely by having Greek and Latin gender endings respectively.

Example: *Coccus* and *Coccos.*

(5) Words having the same meaning and differing only slightly in form.

Example: *cinnamoneus* and *cinnamomeus.*

(6) Words differing only in the presence or suppression with transliteration of diacritic marks (see Rule 65).

Example: *Muellerina* and *Müllerina.*

(7) Words derived from a language other than Latin or Greek, one of which has the ending latinized and the other does not.

Example: *londonensis, london.*

B. Names and Epithets Formed from Personal Names

Generic or Subgeneric Names

(1) When the name for a genus or subgenus is taken from the name of a person, the ending should be formed in the following manner.

TABLE 4. *Endings of personal names for genera and subgenera*

Ending of personal name	Addition	Example
-a	-ea	*Rochalimaea*
Other vowel or -y	-a	*Gaffkya*
-er	-a	*Zinssera*
Other consonant	-ia	*Nocardia*
Latinized name ending in -us	Drop -us, add suffix	*Linnaea* (plant genus)

In addition, the name may be formed by adding a diminutive ending to the name of the person.

Example: *Salmonella, Klebsiella.*

(2) The syllables which are not modified by these endings retain their original spellings, even with the consonants *k* or *w,* or groupings of vowels and consonants not used in classical Latin.

Example: *Wolbachia, Kurthia.*

Epithets

A new specific or subspecific epithet taken from the name of a man or woman may assume (1) an adjectival or (2) a substantival form.

(1) When the epithet is an adjective, it is formed by the addition of an appropriate adjectival suffix, the terminal ending of which agrees in gender with the gender of the genus.

Example: The name "Gordon" takes the adjectival form *gordonianus, gordoniana,* or *gordonianum,* depending on the gender of the generic name and not on whether the name Gordon refers to a male or a female; e.g., *Bacillus gordonian***us**, *Sarcina gordonian***a**, *Propionibacterium gordonian***um**. (These are hypothetical examples.)

(2) When the epithet is a substantive, the terminal ending of the specific epithet varies with the gender of the person to whom the name refers and not with the gender of the generic name. It takes the form of the modernized Latin genitive and is formed in the following manner.

(a) *Greek, Latin or latinized personal names.*

Addition: Appropriate latin genitive.

Example: *Streptococcus pastoris* from *Pastor,* the latinized form of Pasteur.

(b) *Other names.*

(1) *Personal names ending with 'a'. Addition: -i* or *-e* (masculine), *-e* (feminine).

Examples: Name—Shiga (m.) epithet: *shigae;* name—Krzemieniewska (f.) epithet: *krzemieniewskae.*

(2) *Personal names ending in other vowels or -er. Addition: -i* (masculine), *-ae* (feminine).

Examples: Name—Sonne (m.) epithet: *sonnei;* name—Sordelli (m.) epithet: *sordellii;* name—Barker (m.) epithet: *barkeri.*

Note. If the persons named had been females, the endings would have been *-ae* in each case.

(3) *Personal names ending with a consonant (other than -er). Addition: -ii* (masculine), *-ae* (feminine).

Examples: Name—Welch (m.) epithet: *welchii;* name—Gordon (f.) epithet: *gordonae.*

Note. The syllables of the epithet which are not modified by the endings retain their original spellings even with the consonants *k* and *w* or with groupings of vowels or consonants not used in classical Latin.

Appendix 10

Infrasubspecific Subdivisions

Appendix 10. Infrasubspecific Subdivisions

The designations of these taxa are not covered by the Rules of this Code, but this Appendix is included to encourage conformity and to clarify the application of these designations (see Rule 14a, b).

A. Definitions

The term **infrasubspecific subdivision** (or division) has been used in two ways to denote both terms and taxa. It is preferable to distinguish them as given below. **Infrasubspecific "subdivision"** has been used rather than "division" to avoid any confusion with the taxonomic category "division" (*divisio*).

Note. Infrasubspecific subdivisions are not arranged in any order of rank, and may overlap one another.

(1) *Infrasubspecific taxa.* An **infrasubspecific taxon** is one strain or a set of strains showing the same or similar properties, and treated as a taxonomic group.

Example: *Staphylococcus aureus* phagovar 81.

The sets of properties used may be of a similar kind but are not necessarily the same.

Example: The susceptibility to a different phage may be used to define another phagovar of *Staphylococcus aureus*, e.g., phagovar 42D.

Infrasubspecific taxa based on different sets of properties may overlap, e.g., one serovar may contain strains belonging to different phagovars.

Example: *Salmonella typhi* serovars, phagovars, and biovars.

(2) *Infrasubspecific terms.* An **infrasubspecific term** is used to refer to the kinds of taxa below subspecies.

Examples: *serovar, chemovar, forma specialis.*

If a species has not been divided into subspecies, the infrasubspecific terms may be applied to other subdivisions within that species. The subdivisions so named would still be infrasubspecific subdivisions for nomenclatural purposes until such time as they may be raised to subspecific or specific rank.

Example: Serovars of *Erysipelothrix rhusiopathiae.*

(3) *Use of other terms.* **Infrasubspecific form** has been used to refer to a bacterial strain, but this use should be avoided.

A culture of bacteria is a population of bacterial cells in a given place at a given time, e.g., in this test tube or on that agar plate. It may have a longer duration, e.g., desiccated cultures.

A clone is a population of bacterial cells derived from a single parent cell.

A strain is made up of the descendants of a single isolation in pure culture. A strain is usually made up of a succession of cultures and is often derived from a single colony. The number of bacteria which gave rise to the original colony is often unknown. Most bacterial strains are not known to be clones.

Individual is a term with little meaning in bacteriology and has been applied to a single bacterial cell or to a bacterial strain; therefore, it is best to avoid the use of this term.

B. Infrasubspecific Terms

Table 5 contains some of the terms which are commonly used, and the preferred name appears in the first column. The introduction of the suffix **"-var"** or **"-form"** to replace "-type" is recommended to avoid confusion with the strict use of the term "type" to mean nomenclatural type (see Rule 15).

TABLE 5. *Infrasubspecific Terms*

Preferred name	Synonyms	Notes
Biovar	Biotype, physiological type	Biochemical or physiological properties.
Chemoform	Chemotype	Chemical constitution.
Chemovar		Production or amount of production of a particular chemical.
Cultivar		A cultivated strain with special properties.
forma specialis	Special form	A parasitic, symbiotic, or commensal microorganism distinguished primarily by adaptation to a particular host or habitat. Named preferably by the scientific name of the host in the genitive.
Morphovar	Morphotype	Morphological characteristics.
Pathovar	Pathotype	Pathogenic reactions in one or more hosts.
Phagovar	Phagotype, lysotype	Reactions to bacteriophage.
Phase		Restrict to well-defined stages of naturally occurring alternating variations.
Serovar	Serotype	Antigenic characteristics.
State		Colonial variants, e.g., rough, smooth, mucoid (may be defined antigenically).

The term **"type"** in bacteriology should be used strictly for a nomenclatural type (Principle 5 and Chapter 3, Section 4). It should not be used to designate a division of a species nor to designate taxa based on antigenic characters.

The term **"group"** is informal and has no nomenclatural standing. It may prove useful to designate informally a set of organisms having certain characteristics in common, provided that it is used with care and exact definition to avoid ambiguity. It should not be used to avoid the use of the correct name of a taxon such as genus or species.

However, it may be useful when the bacteriologist does not wish to give a formal name to a set of bacteria until further studies have been made but wishes to publish his results and seek the opinion of others.

Example: "IID group," later named *Cardiobacterium hominis.*

C. Nomenclature of Infrasubspecific Taxa

An **infrasubspecific taxon** is designated or cited by the name of the species followed by the infrasubspecific term used to designate this infrasubspecific subdivision followed by the infrasubspecific designation.

Example: *Staphylococcus aureus* phagovar 81.

Reference strains of infrasubspecific taxa may be designated.

There are many ways that infrasubspecific taxa may be designated; among these are the following latinized words, e.g., *cerealis* in *Xanthomonas translucens* f.sp. *cerealis*; vernacular names or words, e.g., rough phase; numbers, letters, or formulae, e.g., phagovar 42D in *Staphylococcus aureus* phagovar 42D.

D. Nomenclature of Strains

A strain may be designated in any manner, e.g., by the name of an individual, by a locality, or by a number.

Statutes of the International Committee on Systematic Bacteriology

Statutes of the International Committee on Systematic Bacteriology

Introduction

The International Committee on Systematic Bacteriology (ICSB) was established by the International Association of Microbiological Societies (IAMS). Statements governing the organization and administration of the Committee and its subordinate bodies were originally contained in Provisions 4 and 5 of earlier editions of the *International Code of Nomenclature of Bacteria (and Viruses)*.

At the ninth International Congress of Microbiology held in Moscow in 1966 the IAMS amended its own Statutes to require that all subordinate bodies, including the ICSB, produce their own Statutes for approval by the Executive Board of IAMS (EBIAMS).

Accordingly, Provisions 4 and 5 of the earlier Codes were deleted from the Code and reconstituted as the Statutes of the ICSB. The first edition of the Statutes was approved by the ICSB at the tenth International Congress of Microbiology in Mexico in 1970 and amended at, and subsequently approved by, the EBIAMS following the first Congress of the Bacteriology Section of IAMS in Jerusalem in 1973.

The EBIAMS also authorized some changes in the statements relating to election of members to the ICSB, the prime purpose of which was to restore to the Executive Secretary of the ICSB the function of communicating directly with Member Societies for nominations of representatives to the ICSB.

An important amendment to the Statutes approved by EBIAMS was that which permits the ICSB to elect members to the Judicial Commission from the membership of the Member Societies instead of solely from the membership of the ICSB. All Judicial Commission members also are automatically members of the ICSB.

The ICSB is so constituted that each Member Society of the IAMS is entitled to appoint one representative. Other members may be coopted. The Committee has an Executive Board, a Judicial Commission on Nomenclature, a Board of Publications, an Editorial Board for the *Inter-*

national Journal of Systematic Bacteriology (IJSB), and several Subcommittees on Taxonomy.

The functions of the ICSB (see Statutes, Article 3) include the following: (i) to hold meetings as part of the sessions of the International Congresses and to sponsor a session on taxonomy at the Congresses; (ii) to approve the *International Code of Nomenclature of Bacteria* and those changes recommended by the Judicial Commission; (iii) to receive and approve the Opinions issued by the Judicial Commission; (iv) to set up special Subcommittees on Taxonomy to study and make recommendations on the classification and nomenclature of bacterial taxa; and (v) to publish the *International Code of Nomenclature of Bacteria,* certain official lists of bacterial names, Statutes of the ICSB, and the IJSB.

The Judicial Commission of the ICSB is a Subcommittee of the ICSB charged with the responsibility of dealing with nomenclatural matters, sometimes in a judicial capacity (see Statutes, Article 8c). The Chairman and Secretaries of that Committee and the Editor of the IJSB are *ex officio* members of the Commission.

The Judicial Commission: (i) considers all proposals for emendation of the *International Code of Nomenclature of Bacteria* and recommends emendations to the ICSB; (ii) establishes lists of names that have been conserved or rejected and other lists of use in bacterial taxonomy; (iii) receives proposals for the establishment of minimal standards for the descriptions of taxa and makes recommendations to the ICSB for approval; (iv) considers Requests for Opinions on the interpretation of the Code and on specific exceptions to the Code (including the conservation or rejection of names of bacterial taxa) in the interest of stability (see Appendix 8 for detailed advice on the preparation of a Request for an Opinion); and (v) considers proposals for the designations of infrasubspecific subdivisions made by the international taxonomic Subcommittees of the ICSB.

V. B. D. SKERMAN
Chairman, International Committee on Systematic Bacteriology

Brisbane, Australia

Statutes of the International Committee on Systematic Bacteriology of the International Association of Microbiological Societies

Article 1

Name

The name of the Committee shall be the International Committee on Systematic Bacteriology of the International Association of Microbiological Societies (IAMS). For purpose of abbreviation in all official documents it will be referred to as the ICSB.

Article 2

Membership on the Committee shall be as Full Members, Coopted Members, and Life Members. Each Member Society (as defined in Article 3 of the Statutes of the IAMS) of the International Association of Microbiological Societies is entitled to appoint one Full Member to the Committee. The ICSB may coopt other members and appoint Life Members.

Full Members. A request for the appointment of a representative to the ICSB will be issued by the Executive Secretary ICSB to the Secretary of each Member Society immediately following each International Congress of Bacteriology and a list of such nominated members forwarded to the Secretary-General IAMS.

In the event of a vacancy arising in the Full Membership during intercongress periods, the Executive Secretary shall request the Member Society to appoint a replacement and shall advise the Secretary-General of such replacements.

The tenure of a Full Member is for one term but he shall be eligible for reelection by his Society for an indefinite period. The nomination must, however, be renewed after each Congress.

A Full Member shall give notice of resignation from the ICSB in writing to the Secretary of the Society which submitted the nomination and to the Executive Secretary of the ICSB.

When a Full Member fails to participate, the Executive Board ICSB (EBICSB) may request the Member Society to nominate a replacement and shall advise the Secretary-General IAMS of such replacements.

Coopted Members. The ICSB may coopt members only from a country which has a Member Society within the terms of Article 3 of the Statutes of the IAMS for the purpose of assisting with the work of the Committee. The Secretary-General IAMS may authorize cooption from other sources.

Proposal of a member for cooption may be made by the EBICSB, by any organization within a country which does not have a Member Society or by a Member Society.

Proposals shall be submitted in writing to the Executive Secretary.

A Coopted Member shall serve until the end of the next Plenary Meeting of the ICSB, shall have the right to vote but shall not be eligible for election to office in the ICSB.

Life Members. The ICSB may appoint to Life Membership of the ICSB, individuals who have rendered distinguished service to the ICSB. Such Life Members shall be considered as Members-at-Large and not as representing the bacteriologists of any Member Society. They shall share all the privileges of membership other than that of election to office in the ICSB.

Each member is entitled to one vote of the Committee. Decisions shall be reached on the basis of a majority of the members present voting yes or no. Abstentions will not be counted. In postal ballots the decision shall be based on the votes received, not on the number of members eligible to vote.

Article 2a

Recognition of Alternates

If a member of the ICSB cannot attend meetings of the ICSB, an alternate having all the rights of an appointed member, except that of eligibility to office in the ICSB, will be appointed in accordance with the following provisions.

(1) The Member Society which the member represents shall have the right to appoint an alternate from within its own society or from within any other Member Society or Institution which is a member of the IAMS.

(2) If no appointment is made by the Member Society, the member himself shall have the right to appoint such an alternate.

(3) If neither the Member Society nor the member appoints an alternate, the Chairman of the ICSB with the Secretaries (Executive and for Subcommittees) may appoint.

Notice of appointment of alternates shall be made to the Executive Secretary of the ICSB and shall not be made to the Secretary-General of IAMS.

All appointments of alternates by Member Societies, or by members, shall be in writing and shall be in the hands of the Executive Secretary, if possible, one month before the next Plenary Meeting of the ICSB.

Article 3

Functions of the ICSB

(1) To hold Plenary Meetings of the ICSB at such times and places as may be determined by the EBICSB, provided that such Plenary Meetings shall be held at intervals of not more than four years and shall, where possible, be held in association with other meetings, conferences, or congresses sponsored by the ICSB or the IAMS.

For the purpose of these Statutes the word "term" shall refer to the period which shall elapse between one Plenary Meeting and the next and shall be of no fixed duration.

(2) To elect from the Full Membership of the ICSB an Executive Board as defined in Article 4.

(3) To elect the members of the Judicial Commission as vacancies occur and to replace the members of the several classes as their terms expire.

(4) To authorize the EBICSB to perform such functions as are defined in Article 4 and such other functions as the ICSB may determine.

(5) To consider and vote upon all recommendations made by the EBICSB relative to matters listed under Functions of the EBICSB (Article 4).

(6) To approve recommendations from the Judicial Commission for the recognition of designations of subdivisions of species and subspecies below the category of subspecies. Such recommendations shall have been made to the Judicial Commission by a Subcommittee on Taxonomy (or other group of specialists appointed by the ICSB) but the Commission shall have the right to conduct such enquiries as it deems necessary before transmitting such recommendations, in their original or amended form, to the ICSB for approval.

(7) To consider and vote upon all recommendations made by the Judicial Commission relating to the *International Code of Nomenclature of Bacteria.*

(8) To consider and vote upon all recommendations submitted by the EBICSB relating to amendments to the Statutes of the ICSB.

(9) To receive all Opinions issued by the Judicial Commission and authorize their publication with the *International Code of Nomenclature of Bacteria* and the Statutes of the ICSB in book form.

(10) To adjudicate, by letter ballot, on issues relating to Requests for Opinions, and appeals against such Requests, as may be referred to it by the Judicial Commission.

(11) To receive Reports of the Subcommittees on Taxonomy.

(12) To establish Publications deemed necessary for the advancement of systematic bacteriology.

(13) To sponsor International Conferences on Systematic Bacteriology, or symposia and paper sessions on Systematic Bacteriology at other conferences sponsored by the IAMS.

(14) To appoint as Life Members of the ICSB, individuals who have rendered distinguished service to the ICSB.

Article 4

The Executive Board of the International Committee on Systematic Bacteriology (EBICSB)

The membership of the Board shall be: The Chairman and Vice-Chairman of the ICSB, the Executive Secretary, the Secretary for Subcommittees, ~~the Editorial Secretary~~, the Treasurer of the ICSB, and two Members-at-Large. The Chairman of the Judicial Commission shall be, *ex officio*, a member of the Executive Board.

Functions of the EBICSB.

(1) To organize Plenary Meetings of the ICSB and prepare and publish agenda for such Meetings.

(2) To conduct the business of the ICSB between Plenary Meetings as directed by the ICSB.

(3) To organize International Conferences and symposia on Systematic Bacteriology.

(4) To appoint Subcommittees on Taxonomy, either on its own initiative, or at the request of others, to approve of the membership of such Subcommittees, and to arrange the schedules for meetings of the Subcommittees at Plenary Meetings of the ICSB.

(5) To provide for the initial appointment and subsequent election of Chairmen and Secretaries of Subcommittees on Taxonomy.

(6) To deal with such business as the Judicial Commission may from time to time request.

(7) To recommend to the ICSB the establishment of any Committees that are deemed necessary for the work and functioning of the ICSB and to consider recommendations submitted to it by those Committees.

(8) Upon approval of the ICSB to:

(a) establish Publications authorized by the ICSB.

(b) negotiate such contracts as may be necessary for the issuance of Publications authorized by the ICSB;

(c) establish such Trusts or enter into such agreements as may be advisable for the auditing and administration of funds which may be designated for the payment of the necessary operating expenses of the ICSB and its subordinate agencies, whether such funds originate from other grants, gifts, royalties, the sale of publications, or other sources.

(9) To request from appropriate agencies grants for the necessary expenses of the ICSB and its subordinate agencies.

(10) To ensure the proper functioning of the organizations and officers of the ICSB.

(11) To administer Article 15 (Amendment to the Statutes).

(12) To perform such other functions assigned to it by the ICSB.

Article 5

Election and Duties of Chairman and Vice-Chairman of the ICSB

At each Plenary Meeting, the ICSB shall elect a Chairman and a Vice-Chairman, who shall hold office until the close of the next Plenary Meeting, and shall be eligible for re-election for a maximum of two terms. They shall be *ex officio,* but non-voting, members of all Committees and Subcommittees of the ICSB, unless otherwise stated.

A. Duties of the Chairman of the ICSB.

(1) To preside at meetings of the ICSB and its Executive Board.

(2) With the Secretaries, to prepare agenda for meetings of the Executive Board of the ICSB.

(3) With the other members of the Executive Board to prepare agenda for meetings of the ICSB.

(4) To receive and review Reports and Minutes from the Subcommittees on Taxonomy and in collaboration with the Editorial Secretary prepare statements for publication, with the approval of the Chairmen of the Subcommittees, in the *International Journal of Systematic Bacteriology*.

(5) To assume such duties as may be requested by the ICSB, or determined by the Functions of the EBICSB.

B. Duties of the Vice-Chairman.

(1) To preside at the meetings of the ICSB and EBICSB in the absence of the Chairman.

(2) To preside at meetings of the Publications Committee as its Chairman.

(3) With the other members of the Executive Board to prepare agenda for the meetings of the ICSB.

(4) To assume such duties as may be requested by the ICSB or determined by the Functions of the EBICSB.

Article 6

Election of Secretaries

There shall be three Secretaries of the ICSB, elected by the ICSB from its membership, or from the membership of a Member Society. If not a member of the ICSB at the time of election to office, a Secretary shall be, *ex officio,* a member of the ICSB during his term of office with all the rights and privileges of membership.

The Secretaries shall be designated:

(1) Executive Secretary.

(2) Secretary for Subcommittees on Taxonomy.

(3) Editorial Secretary.

Nominations for the filling of a vacancy in the position of a Secretary which may occur between Plenary Meetings of the ICSB shall be made by the EBICSB, and submitted to the ICSB for postal ballot immediately or submitted to a vote of the ICSB at the next Plenary Meeting.

The Secretaries shall be elected to hold office for two consecutive terms and shall not be eligible for re-election without the prior assent of the Secretary-General of IAMS. Automatic retirement of the Secretaries shall be in rotation, no more than two retiring at any Plenary Meeting of the ICSB.

The Secretaries will serve *ex officio* as voting members of the Executive Board of the ICSB, the Judicial Commission, the Publications Committee and its Editorial Boards, and the ICSB.

Article 6a

Duties of the Executive Secretary

The Executive Secretary shall be responsible to the Executive Board in all matters associated with communications between the ICSB and the EBIAMS and Member Societies. He shall perform the following duties.

(1) Be Secretary of the EBICSB and the ICSB and submit the Minutes of their meetings to the Editorial Secretary for publication in the *International Journal of Systematic Bacteriology*.

(2) Be Secretary of the Publications Committee.

(3) Prepare, in cooperation with the Chairman of the ICSB and the Secretary for Subcommittees, agenda for the meetings of the EBICSB.

(4) Prepare, in cooperation with other members of the EBICSB, agenda for all meetings of the ICSB.

(5) Prepare, in cooperation with other members of the Publications Committee, agenda for all meetings of the Publications Committee.

(6) Secure from each Member Society their nomination of a representative to the ICSB and welcome and advise such members of their duties as members of the ICSB. He shall circulate the List of Members to the EBICSB and to the Secretary-General of IAMS.

(7) Receive from the Secretaries of Member Societies, or from members of the ICSB, nominations for alternates for meetings of the ICSB and submit a list of these to the EBICSB before the commencement of each meeting.

(8) Prepare at the conclusion of each Plenary Meeting of the ICSB a complete list of all members and officers of the ICSB and Judicial Commission and forward these to the Secretary-General IAMS, the Chairman of ICSB, the Chairman of the Judicial Commission, the Editorial Secretary, and the Editors of all publications approved by the ICSB.

(9) Receive proposals for appointment of coopted and Life Members and transmit these to the EBICSB.

(10) Transmit to the ICSB from the Judicial Commission such recommendations as may require action by the ICSB. If the members of the ICSB are circularized, to secure comments and suggestions, to tabulate the information received. If the members are asked to vote upon any proposal, to tabulate and announce the result of the ballot, and to certify

the result to the Chairman of the ICSB and to the Chairman of the Judicial Commission.

(11) Present to the Plenary Meeting of the ICSB a Report covering all pertinent actions of the ICSB and its Judicial Commission and to forward a copy of the Report to the Secretary-General of the IAMS.

(12) With the other members of the EBICSB, authorize the formation of new subcommittees to consider the taxonomy of special groups of microorganisms.

(13) Perform such duties as the EBICSB may from time to time determine.

Article 6b

Duties of the Secretary for Subcommittees on Taxonomy

The Secretary for Subcommittees on Taxonomy shall be responsible to the Executive Board for all matters associated with the Subcommittees on Taxonomy of the ICSB. He shall perform the following duties.

(1) Prepare, in cooperation with the Chairman ICSB and the Executive Secretary, agenda for meetings of the EBICSB.

(2) Prepare, in cooperation with other members of the EBICSB, agenda for meetings of the ICSB.

(3) Act as Liaison Secretary to all Subcommittees on Taxonomy and transmit their Reports and Minutes to the Editorial Secretary and Requests and Recommendations to the EBICSB or Judicial Commission, whichever is appropriate.

(4) With the other members of the EBICSB, authorize the formation of new Subcommittees to consider the taxonomy of special groups of microorganisms.

(5) Maintain lists of the officers and members of the Subcommittees on Taxonomy and to transmit these at the conclusion of each Plenary Meeting of the ICSB to the Secretary-General of IAMS, the Chairman of the ICSB, the Executive Secretary, and the Editorial Secretary for publication in the IJSB.

Article 6c

Duties of the Editorial Secretary

The Editorial Secretary shall be Secretary to the Judicial Commission and to the Editorial Board for the *International Code of Nomenclature of Bacteria.* He shall perform the following duties.

(1) Prepare, with the Chairman of the Judicial Commission, agenda for meetings of the Judicial Commission.

(2) Prepare, with the Chairman of the Editorial Board for the *International Code of Nomenclature of Bacteria,* agenda for meetings of that Board.

(3) Prepare and submit copies of the Minutes of meetings of the Judicial Commission to the Executive Secretary together with any formal statements of Opinions or Recommendations for transmission by him to the EBICSB and ICSB and to the Editor of the IJSB for publication.

(4) Receive all requests for Opinions or for assistance on nomenclatural problems and to refer these to the Chairman of the Judicial Commission and, where necessary, to the Editorial Board of the *International Code of Nomenclature of Bacteria.*

(5) Submit to the Editor of the *International Journal of Systematic Bacteriology* Opinions of the Judicial Commission for publication.

(6) Receive Minutes and Reports of the Subcommittees on Taxonomy from the Secretary for Subcommittees and transmit them to the Chairman of the ICSB, and collaborate with the Chairman of the ICSB in the preparation of Statements from the Reports and Minutes of the Subcommittees on Taxonomy and to submit them, after approval by the Chairman of the Subcommittee, to the Editor of the IJSB for publication.

(7) Receive proposals for emendation of, and to assist in the preparation of, all documents relating to the *International Code of Nomenclature of Bacteria* for consideration by the Judicial Commission.

(8) Receive all other submissions for publication in the IJSB and transmit them, with or without prior editing, to the Editor of that Journal.

Article 7

Election of Treasurer

The office of Treasurer of the ICSB may be combined with that of the Executive Secretary on the recommendation of the EBICSB. The Treasurer shall be elected by the ICSB. The Treasurer shall be elected to hold office for two consecutive terms and shall not be eligible for re-election without the prior assent of the Secretary-General IAMS.

Duties of the Treasurer

The Treasurer shall receive funds which may be made available to the ICSB from any sources, and distribute them as directed by the EBICSB or its Chairman.

Cheques issued by the Treasurer shall bear his signature only, but must have been authorized by the Chairman of ICSB.

An annual statement of accounts shall be furnished by the Treasurer to the Executive Secretary (if the Treasurer does not hold that office) and an audited statement shall be furnished to the EBICSB and the ICSB and the Secretary-General of IAMS at each Plenary Meeting of the ICSB.

Article 8

The Judicial Commission

Organization. The Judicial Commission shall consist of seventeen members, twelve elected by the members of the ICSB, the Chairman of the ICSB, and the three Secretaries. The Editorial Secretary of the ICSB is also the Secretary of the Judicial Commission. The Editor of the IJSB shall be, *ex officio,* a member of the Judicial Commission.

Commissioners are elected to serve in three classes of four Commissioners each, one class retiring at the close of each Plenary Meeting of the ICSB. Members of retiring classes are eligible for re-election.

All Commissioners shall be accorded membership on the ICSB with power to vote.

In the case of resignation or death of a Commissioner between Plenary Meetings, the vacancy may be filled by letter ballot of the members of the ICSB, after the vacancy on the ICSB has been filled by the Member Society. A Commissioner of one Class is eligible for election to a vacancy occurring in another Class. Commissioners elected to fill a vacancy caused by resignation or death shall serve for the unexpired term of the vacancy.

If any Commissioner cannot attend the meetings of the Judicial Commission, an alternate having all the rights of a Commissioner except in the election of officers will be chosen in accordance with the following provisions.

(1) The Commissioner himself shall name an alternate from outside the Commission, but from the members of the ICSB.

(2) If no name is submitted by the Commissioner, the Judicial Commission may name an alternate from members of the ICSB attending the Plenary meeting.

(3) No alternate shall represent more than one absent Commissioner.

The names of all alternates [(1) and (2) above] shall be in writing and in the hands of the Secretary to the Judicial Commission before the commencement of the first of the meetings of the Judicial Commission scheduled at any Plenary Meeting of the ICSB, and they shall be presented at that meeting.

One of the Commissioners shall be chosen as Chairman and one as Vice-Chairman by vote of the Judicial Commission. The Chairman and Vice-Chairman shall hold office during their unexpired terms as Commissioners, but shall be eligible for re-election if re-elected as Commissioners.

Article 8a

Duties of Chairman of the Judicial Commission

(1) To preside at meetings of the Judicial Commission.

(2) To prepare, with the collaboration of the Editorial Secretary of ICSB, agenda for meetings of the Judicial Commission.

(3) To appoint such committees as are authorized by the Judicial Commission but whose appointment has not been otherwise provided for.

(4) To serve as Chairman of the Editorial Board for the *International Code of Nomenclature of Bacteria.*

(5) To receive Requests for Opinions through the Editorial Secretary; to edit the Requests and assist in the preparation of a statement embodying the edited Request, the form of the Opinion and a commentary (if desired) for publication in the *International Journal of Systematic Bacteriology* and for transmission through the Secretary for Subcommittees to appropriate Subcommittees on Taxonomy.

After a period of six months following the date of publication of the Request, to submit the Request, together with any appeals against the issuance of the Opinion, to the Judicial Commission for vote. In the event of an affirmative vote, to arrange for publication of the Opinion in the *International Journal of Systematic Bacteriology,* or, in the event of the Judicial Commission failing to reach a decision when an appeal has been lodged, to refer the case to the ICSB for determination.

(6) To receive from the Editorial Secretary requests or suggestions for the emendation of the *International Code of Nomenclature of Bacteria;* with other members of the Editorial Board for the Code, to formulate the amendments and to circulate them to the members of the Judicial Commission, which shall make appropriate recommendations to the ICSB for vote.

(7) To represent the Judicial Commission on such International Committees, Boards or Commissions as may be organized to consider cooperation in biology in the solution of common problems of nomenclature and taxonomy, particularly to work with other similar Commissions or Executive Committees organized for action on problems on nomenclature in botany, in zoology, and in virology.

(8) To undertake such other duties as may from time to time be requested by the Judicial Commission.

Article 8b

Duties of Vice-Chairman of the Judicial Commission

(1) To preside at meetings of the Judicial Commission in the absence of the Chairman.

(2) To assume all the duties of the Chairman of the Commission in the event of death or resignation of the Chairman, until such time as a new Chairman has been elected.

Article 8c

Functions of the Judicial Commission

The Judicial Commission has the following functions.

(1) To hold such sessions as may be necessary to transact business.

(2) To consider all Requests for Opinions relative to the interpretation of the Principles, Rules, and Recommendations of the *International Code of Nomenclature of Bacteria* where applications are doubtful.

An Opinion shall be issued when ten or more Commissioners vote in favour of it. All Opinions shall be reported to the ICSB, and unless rescinded by a majority of those voting in this Committee, such Opinions shall be considered final.

(3) To consider each proposal for emendation of the *International Code of Nomenclature of Bacteria*. When approved by ten or more Commissioners, the proposal shall be submitted to the members of the ICSB and shall be considered as approved when it has been accepted by at least seventy per cent of the members voting. Emendations shall be reported to the next Plenary Meeting of the ICSB.

(4) To make recommendations to the ICSB relative to the official designation of subdivisions of species and subspecies below the category of subspecies and to request revision of these recommendations when necessary. For this purpose the Judicial Commission shall invite proposals from the relevant Subcommittee on Taxonomy, or in the absence of such a Subcommittee from an *ad hoc* committee of experts appointed to submit such proposals.

(5) To review lists prepared by the Editorial Board for the *International Code of Nomenclature of Bacteria* of the:

(a) names of taxa that have been conserved (*nomina conservanda*) or rejected (*nomina rejicienda*) on the basis of Opinions issued relative to the status of such names;

(b) types which have been fixed through issuance of Opinions of the Judicial Commission.

(c) names of genera of bacteria that have been validly published and, if found advisable, lists of the generic names of other groups in which microbiologists are interested. Such lists would be designed to assist authors publishing new names and combinations to avoid proposing illegitimate later homonyms.

(d) publications in the field of bacteriology in which names proposed shall be regarded as not validly published and having no standing in bacteriological nomenclature.

(6) To report through the EBICSB to the ICSB at its first session at each Plenary Meeting the names of all Commissioners whose terms of service expire at the close of the Meeting and a list of other vacancies

in the membership of the Commission, all of which should be filled by election by the ICSB.

(7) Through its Chairman and with the collaboration of the Secretaries to cooperate with other Commissions or similar bodies appointed by the International Botanical and Zoological Congresses to consider problems of nomenclature.

(8) To request Subcommittees on Taxonomy, or, in their absence, other specialists, to suggest minimal standards for the description of new taxa and to make recommendations to the ICSB regarding the acceptance of such standards in assessing the validity of a publication for the recognition of new taxa, to publish such minimal standards as may be approved by the ICSB and to maintain a regular review of such standards as circumstances may require. These minimal standards shall in no way limit the extent of further enquiry or the search for alternate bases for setting such standards.

(9) To consider recommendations from Subcommittees on Taxonomy, or other specialists, for the acceptance of a list of names as valid and applicable to recognizable taxa. To further consider recommendations for the relegation of all other names to a list from which names may be reused for application to new taxa.

Article 9

Organization and Functions of Subcommittees on Taxonomy

The formation of a Subcommittee on Taxonomy may be proposed by an individual or by a group of individuals or by the EBICSB. A proposal must be accompanied by a list of the taxa to be studied, the reasons therefore, and a list of proposed members. Such proposals shall be lodged with the Secretary for Subcommittees. Subcommittees on Taxonomy appointed by the EBICSB shall work under the following rules.

(1) *Members of a Subcommittee.* The Chairman of the ICSB, with the other members of the Executive Board, will select the members of the new Subcommittee on Taxonomy from names submitted and other names. They will also appoint a Chairman and Secretary to carry on the work with the members of the Subcommittee until the next Plenary Meeting of the ICSB, at which time the Subcommittee will hold its first election of officers.

(2) New members may be elected to a Subcommittee at any time. Nominations shall be called by the Secretary of the Subcommittee and any nominations received discussed with the Secretary for Subcommittees on Taxonomy *before* they are submitted to the existing members of the Subcommittee for a vote. The Secretary for Subcommittees will conduct an election for ordinary membership only if requested to do so by the

Chairman of the Subcommittee (or, in the non-existence of a Chairman, the Secretary of the Subcommittee) or by the EBICSB.

(3) At times other than at the establishment of a new Subcommittee, the chairman and secretary of each Subcommittee on Taxonomy shall hold office for two consecutive terms. They shall be eligible for re-election. The Secretary for Subcommittees on Taxonomy shall arrange for the election of a Chairman and a Secretary:

(a) when requested to do so by either the Secretary or the Chairman when the other office falls vacant; or

(b) when requested to do so by members of the Subcommittee when both offices fall vacant simultaneously.

(4) The Secretary for Subcommittees on Taxonomy shall be designated as a non-voting member of each Subcommittee and shall act as liaison between the Subcommittee, the EBICSB, and the Judicial Commission.

(5) Each Subcommittee on Taxonomy shall meet at least twice at each Plenary Meeting of the ICSB. One of these meetings may be restricted to members of the Subcommittee or their alternates and shall deal, *inter alia,* with changes of membership and election of officers. The names of members who have resigned or who have ceased to interest themselves in the work of the Subcommittee shall be deleted from the membership list and a revised list of officers and members supplied to the Secretary for Subcommittees before the conclusion of the Plenary Meeting.

The second and subsequent meetings of the Subcommittee shall deal with taxonomic matters and shall be open to others who may, at the discretion of the Chairman of the Subcommittee, participate in discussions but who will have no voting rights.

Subcommittees may hold additional meetings at times other than at Plenary Meetings.

(6) The Chairman and Secretary of each Subcommittee on Taxonomy shall be responsible for the preparation of a report on the activities of the Subcommittee during the period from the closing of one Plenary Meeting of the ICSB up to six months prior to the opening of the succeeding Plenary Meeting. This report will be submitted in writing through the Secretary for Subcommittees on Taxonomy to the EBICSB for transmission to the ICSB. These reports will be read by the Secretaries of the respective Subcommittees at the first session of the next Plenary Meeting of the ICSB. Within one month of the conclusion of the Plenary Meeting, the Minutes of the Meetings of the Subcommittees held at the time of the Plenary Meeting shall be submitted, with the report, to the Secretary for Subcommittees on Taxonomy for transmission to the Chairman of ICSB and to the Editorial Secretary. They shall edit the Reports and Minutes of each Subcommittee and prepare a statement

which, after approval by the Chairman of the Subcommittee, shall be forwarded to the Editor of the *International Journal of Systematic Bacteriology* (IJSB) for publication. Alternatively, the Chairman of the Subcommittee may prepare such a statement for submission through the Secretary for Subcommittees on Taxonomy to the Chairman of ICSB for review and publication.

Note. As a result of studies, Subcommittees may make recommendations regarding (*inter alia*) taxonomic procedures, changes in nomenclature, recognition of types of various taxa, or classification. While statements relating to these may appear in Reports and Minutes, they should also be submitted as separate papers under the title of "Recommendations of the Subcommittee on Taxonomy of . . . .," for publication in the IJSB (for further notes on such submissions see Clause 8 below, Functions of Subcommittees).

(7) A Subcommittee on Taxonomy may coopt non-voting collaborators to study particular problems.

(8) Functions of Subcommittees.

(a) To encourage and to undertake research on the relationships of the organisms in the taxa under study.

(b) To use any or all of the techniques of the several branches of science in the recognition of characters useful in distinguishing the bacteria under study and to make recommendations regarding procedures.

(c) To make recommendations in relation to the classification of the taxon under study. Although a Subcommittee cannot legislate on classification the status of the Subcommittee may contribute materially towards the general acceptance of a classification.

(d) To make recommendations in relation to the nomenclature of the organisms in the taxon under study. This would include recommendations for changes in names and the conservation and rejection of names. It is inadvisable for Subcommittees *per se* to make proposals for new taxa. This should be done in the form of a scientific paper issued under the authorship of preferably not more than two members of the Subcommittee acting as individuals (not as Subcommittee members). When the proposal has been published the Subcommittee should, if it so desires, publish a note in the *International Journal of Systematic Bacteriology* endorsing the views of the authors. If the paper is itself published in the IJSB, citation of endorsement by the Subcommittee, inserted under the title and authorship of the paper, would be adequate.

(e) To offer advice upon, and to request the Judicial Commission to adjudicate between, conflicting proposals for type species of genera and neotype strains of species and subspecies.

Note. Such advice and requests are appropriately made by a Subcommittee as an official body. The decision is ultimately made by the

Judicial Commission which may issue its findings in the form of an Opinion.

(f) To study the *International Code of Nomenclature of Bacteria* and make recommendations to the Judicial Commission regarding emendations. Such recommendations may be made at any time but preferably not less than six months prior to any Plenary Meeting of the ICSB to permit publication in the IJSB before that meeting. The Editorial Secretary will advise on pertinent procedure.

(g) To recommend to the ICSB, through the Judicial Commission, the official designation of subdivisions of species and subspecies below the category of subspecies, e.g. phagovars of *Staphylococcus aureus*.

(h) To recommend to the ICSB through the Judicial Commission minimal standards for the description of new taxa for the purpose of establishing validity of publication. Such recommendations shall include a list of characters and methods for their assessment, and shall be reviewed, at the request of the Judicial Commission, at regular intervals. If accepted by the ICSB, they shall be published in the IJSB and other microbiological journals. They shall specify the minimal requirements only and shall in no way limit the extent of investigation beyond these limits. The Judicial Commission may, at the request of any specialist in the field of study whether he be a member of the Subcommittee or not, call for a revision of the minimal standards if the evidence before the Commission is considered sufficient to warrant such a call.

(i) To:

(*1*) prepare a list of names which have been applied to the taxon under consideration;

(*2*) prepare from (*1*) a list of such names which are considered to be validly published and applicable to recognizable taxa;

(*3*) recommend to the ICSB, through the Judicial Commission, the relegation of all names from (*1*) which are not included in (*2*) to a list from which names may be reused for the naming of new taxa.

(9) Members who cannot attend meetings of Subcommittees may designate alternates to act for them. An alternate shall be provided by the member with written authority to record a vote on the member's behalf.

Article 10

Publications Committee: Organization and Functions

The members of the Publications Committee shall be the Chairman and Vice-Chairman of the ICSB, the Chairman of the Judicial Commission, the Editors of Publications authorized by the ICSB, and the three Secretaries.

The Vice-Chairman of the ICSB shall be Chairman of the Publications Committee.

The Executive Secretary shall be Secretary of the Publications Committee.

Functions of the Publications Committee

The Publications Committee shall:

(1) make recommendations to the ICSB through the EBICSB for the establishment of such publications as are deemed necessary for the advancement of Systematic Bacteriology and Editorial Boards for such publications and coordinate the activities of those Boards;

(2) serve as the Editorial Board for the Statutes of the ICSB and for any publications for which there are no specifically appointed Editorial Boards;

(3) be responsible for the publication of the *International Code of Nomenclature of Bacteria* and the *Statutes of the International Committee on Systematic Bacteriology* in book form;

(4) prepare an Annual Report and Financial Statement relating to publications for transmission through the Editorial Secretary to the EBICSB;

(5) submit a Report of the Publications Committee through the Editorial Secretary to the ICSB at each Plenary Meeting of the ICSB.

Article 11

Editorial Boards

Editorial Boards shall be proposed by the Publications Committee and appointed by the EBICSB.

The Editorial Secretary of the ICSB shall be Secretary to the Editorial Boards established by the EBICSB.

Until otherwise determined two Editorial Boards shall be recognized:

(1) The Editorial Board for the *International Code of Nomenclature of Bacteria.*

(2) The Editorial Board for the *International Journal of Systematic Bacteriology.*

Article 11a

(1) The Editorial Board for the *International Code of Nomenclature of Bacteria* shall consist of an Editor, appointed by the EBICSB, the Chairman of the Judicial Commission, who shall be Chairman of the Board, and the Editorial Secretary. It shall have power to coopt.

(2) The Editor shall be responsible for the continuing revision of the Code and for its editing. He shall submit the manuscript after approval

by the Board to the Secretary of the Publications Committee for publication with the Statutes of the ICSB in book form.

(3) The Board may receive requests for emendations to the Code and shall make such recommendations as are deemed necessary to the Judicial Commission for consideration.

(4) The Board shall prepare the lists cited in Article 8c(5) (a-d) and submit them to the Judicial Commission for review and publication.

(5) The Board shall act as advisory body to the Chairman of the Judicial Commission on the interpretation of the Code.

Article 11b

The Editorial Board for the International Journal of Systematic Bacteriology

(1) This Board shall consist of an Editor elected by the EBICSB, the Editorial Secretary, the Secretary for Subcommittees on Taxonomy, the Chairman of the ICSB, who shall be Chairman of the Board, and the Managing Editor.

(2) The Editor of the Journal shall receive, through the Managing Editor, all material submitted for publication in the *International Journal of Systematic Bacteriology*. He shall have the power to refer such submissions to referees for assessment and reject such submissions which are deemed unsuitable for publication, provided that, in such instances, authors shall have the right of appeal to the Board. He shall refer all submissions dealing with nomenclatural matters to the Chairman of the Judicial Commission for comment before publication. The Managing Editor shall submit copies of all manuscripts received to the Editorial Secretary.

(3) The Editor shall act on behalf of the Board, except in respect to the negotiation of contracts, in all transactions (including financial transactions) with the organization contracted by the IAMS, but he shall forward copies of correspondence dealing with such transactions immediately to other members of the Board. He shall submit an annual report and an audited financial statement to the Board.

(4) The Editor shall make recommendations to the Board relating to any matters connected with publication of the Journal.

(5) The Editorial Board shall make application to the EBICSB for funds which shall be placed at the disposal of the Editor of the IJSB for the purpose of administering the publication of the Journal.

Article 12

Official Publications of the ICSB

The official publications of the ICSB are the *International Journal of Systematic Bacteriology,* and the *International Code of Nomenclature of Bacteria and Statutes of the International Committee on Systematic Bacteriology.*

Article 13

No one connected with a commercial firm may use his connection with the ICSB or any of its organizations, either as a member or an officer, to advertise or promote his firm in any way.

Article 14

No publication of the ICSB or any of its organizations shall bear any indication of sponsorship by a commercial company, or institution connected in any way with a commercial company, except an acceptable acknowledgment of financial assistance.

Article 15

Amendments

Amendments to the Statutes of the ICSB shall be proposed formally only at Plenary Meetings of the ICSB.

Requests for proposals shall be forwarded to members of the ICSB twelve months in advance of the date of the next Plenary Meeting of the ICSB by the Executive Secretary.

A List of the proposed amendments together with a commentary and a ballot paper shall be forwarded to each member of the ICSB before the meeting at a time which will enable the ballot papers to be returned before the meeting.

Those amendments for which there is unanimous acceptance amongst ballots returned will be proposed *en bloc* for formal approval at the ICSB meeting.

Amendments for which there has not been unanimous support will be presented at the meeting and submitted for discussion and a ballot by members attending the meeting. The rules of voting indicated under Article 2 will apply. The decision will be final.

Notice of amendments approved by the ICSB shall be published in the IJSB.

The amended Statutes will be published with the *International Code of Nomenclature of Bacteria* in book form at times to be determined by the Publications Committee and the Publishers.

Statutes of the Bacteriology Section of the International Association of Microbiological Societies

Statutes of the Bacteriology Section of the International Association of Microbiological Societies

STATUTES[1]

Article 1

In accordance with the Statutes of the International Association of Microbiological Societies (IAMS), the name of the Section is "Bacteriology Section." Those articles of the IAMS Statutes referring to the constitutions and activities of subordinate bodies shall apply to this Section.

Article 2

Objectives

The objectives of the Section are:

(a) To maintain contact among bacteriologists and microbiological societies throughout the world within the organization of IAMS through sponsorship of international meetings (congresses, symposia, conferences, and colloquia), distribution of current information in the field of bacteriology, and support of other appropriate activities.

(b) To maintain contact with other Sections of IAMS through exchange of information on Section activities and holding Intersectional Meetings in accordance with rulings of IAMS.

(c) To be responsible to the EBIAMS in accordance with Articles 8F, 8G, and 8H of the Statutes of IAMS.

(d) To support Committees, Federations, and Commissions as may be appropriate for international cooperation and agreement.

(e) To represent bacteriologists as a Section in IAMS.

(f) To encourage research in bacteriology.

(g) To encourage the highest standard of training of bacteriologists of all nations so that bacteriology may be used for the fullest benefit of mankind.

[1] Approved (unanimously) by the Statutes Committee, October 1971, and Amended by the Section Council, 2 September 1973, Jerusalem, Israel, and submitted by R. R. Colwell, Secretary-Treasurer, Bacteriology Section, International Association of Microbiological Societies.

Article 3

Membership

Member Societies belonging to IAMS and having an active interest in Bacteriology may become Member Societies of the Section. A country may have more than one Member Society in the Section.

Article 4

Section Council

(A) The Section Council shall be composed of:

1. Delegates from each Member Society interested in the fields of the Section. The number of these delegates is to be determined by Article 6(A) of the IAMS Statutes, each nation represented to have one vote. In nations having more than one Member Society, the National Committee shall designate the voting delegate (or the delegate with the power to vote). (Article 3, paragraph 2, of the IAMS Statutes.) The other national delegates may speak at the Council Meetings and shall be kept informed of, and consulted on, all matters submitted to the Section Council.

2. The Chairman, or his designate, of each International Committee, Federation or Commission interested in the field of the Section, who may speak at Council Meetings and shall be kept informed of, and consulted on, all matters submitted to the Section Council, but shall not be a voting delegate.

(B) The Section shall choose officers in accordance with Article 8(B) of the IAMS Statutes.

1. The officers of the Section shall consist of a Chairman, a Vice-Chairman, and a Secretary-Treasurer. They shall be elected by all the voting delegates to the Section Council. These 3 officers plus the immediate Past-Chairman shall constitute the Executive Committee of the Section.

2. Should a vacancy on the Executive Committee of the Section Council occur between meetings by reason of health or resignation, nomination for replacement may be made by the remaining members of the Executive Committee. The Secretary-Treasurer shall conduct an election by mail ballot of the Section Council, a majority vote of members shall constitute election. Votes shall be recorded within a time set by the Secretary-Treasurer in each case.

3. The main duty of the Executive Committee is to organize at intervals of not less than 4 years an International Congress for Bacteriology and, between congresses, such other meetings (symposia, conferences, colloquia) as may be appropriate.

(1) The time and place of the next Congress will be decided by

the Executive Committee in consultation with the Council at one Congress in anticipation of the next.

(2) For each Congress, there shall be a Program Committee composed of a Convenor, appointed by the officers of the Bacteriology Section, and the Bacteriology Section Council. Local and program arrangements for the Congress shall be in the hands of a local Host Committee. The program shall be submitted for approval by the Section Council.

4. The duties of the Chairman shall be:

(1) To preside at meetings of the Executive Committee and Council of the Section.

(2) To prepare with the Secretary-Treasurer the agenda for meetings of the Executive Committee of the Section.

(3) To prepare with the Secretary-Treasurer and other members of the Executive Committee the agenda for meetings of the Section Council.

(4) To attend the meetings of the Executive Board of IAMS.

5. The duties of the Vice-Chairman shall be:

(1) To preside at the meetings of the Executive Committee and Council of the Section in the absence of the Chairman.

(2) To prepare with other members of the Executive Committee the agenda for meetings of the Section Council.

(3) To attend the meetings of the Executive Board of IAMS in the absence of the Chairman.

6. The duties of the Secretary-Treasurer shall be:

(1) To prepare with the Chairman agenda for meetings of the Executive Committee of the Section.

(2) To prepare with other members of the Executive Committee agenda for meetings of the Section Council.

(3) To attend the meetings of the Executive Board of IAMS in the absence of the Chairman and Vice-Chairman.

(4) To present to the Plenary meeting of the Section Council a report covering all pertinent activities of the Executive Committee.

(5) To correspond with the Member Societies and to perform duties as the Executive Committee of the Section may from time to time determine.

(6) To handle any funds that may be allocated to the Section by IAMS or received from other sources approved by IAMS.

(7) To conduct elections and other duties that are necessary to carry on the business of the Section.

(C) The duties of the Officers and Council.

Amongst other things they shall:

1. Organize the Section effectively.

2. Maintain communication with:

(a) The Executive Board of IAMS,

(b) Other Sections of IAMS, and

(c) Existing Committees, Federations and Commissions that are relevant to the Section.

3. Form committees as necessary for carrying out the duties and functions of the Section.

4. Hold plenary meetings of the Section Council.

5. Elect from the membership of the Council the Executive Committee (Chairman, Vice-Chairman, and Secretary-Treasurer).

6. Consider and vote upon the recommendations of the Executive Committee relative to the activities of the Section.

(D) The Secretary-Treasurer of the Section shall provide the Secretary-General of IAMS and the Secretaries of the other Sections, International Committees, Federations, and Commissions with proposals and notices of congresses, symposia, conferences, colloquia, or other meetings.

(E) The Section Council shall meet at the time of Section and Intersection meetings.

Article 5

No one connected with a commercial firm may use his connection with the Bacteriology Section, or Committee of the Section, either as a member or officer, to advertise or promote his company in any way.

Article 6

No publication of the Section or Committee of the Section shall bear any indication of sponsorship by a commercial company or institution connected in any way with a commercial company, except an acceptable acknowledgment of financial assistance. Furthermore, any publication containing material not authorized, prepared, or edited by the Section or Committee of the Section may not bear the name of IAMS without specific permission of the EBIAMS.

Article 7

Disposition of any remaining funds: in the event of dissolution of the Section any remaining funds shall be turned back to the Treasurer of IAMS.

Article 8

The Statutes of the Section may be amended by a two-thirds majority vote of the voting delegates present at a General Meeting. Any Section Council member may propose amendments which are to be sent to the

Secretary of the Section and transmitted to the Section Council members, at least six months prior to the General Meeting of the Section Council. The amendments must afterwards be approved by the Executive Board of the IAMS.

Article 9

The legal seat of the Section shall be that of the Secretary-Treasurer.

Index

and References to Definitions

The following terms are defined in the clauses of the Code to which reference is given. Additional information may be found in S. T. Cowan. 1968. *A Dictionary of Microbial Taxonomic Usage*. Oliver and Boyd, Edinburgh. 118 pp.

Author Index

Subject Index